Legal Notice

This book is copyright 2019 with all rights reserved. It is illegal to copy, distribute, or create derivative works from this book in whole or in part or to contribute to the copying, distribution, or creating of derivative works of this book.

For information on bulk purchases and licensing agreements, please email

support@SATPrepGet800.com

ISBN-13: 978-1-951619-91-6

This is the Solution Guide to the book "Pure Mathematics for Beginners."

Also Available from Dr. Steve Warner

CONNECT WITH DR. STEVE WARNER

www.facebook.com/SATPrepGet800

www.youtube.com/TheSATMathPrep

www.twitter.com/SATPrepGet800

www.linkedin.com/in/DrSteveWarner

www.pinterest.com/SATPrepGet800

Also Available from Dr. Steve Warner

CONNECT WITH DR. STEVE WARNER

www.facebook.com/SATPrepGet800

www.youtube.com/TheSATMathPrep

www.twitter.com/SATPrepGet800

www.linkedin.com/in/DrSteveWarner

www.pinterest.com/SATPrepGet800

Pure Mathematics for Beginners

Solution Guide

Dr. Steve Warner

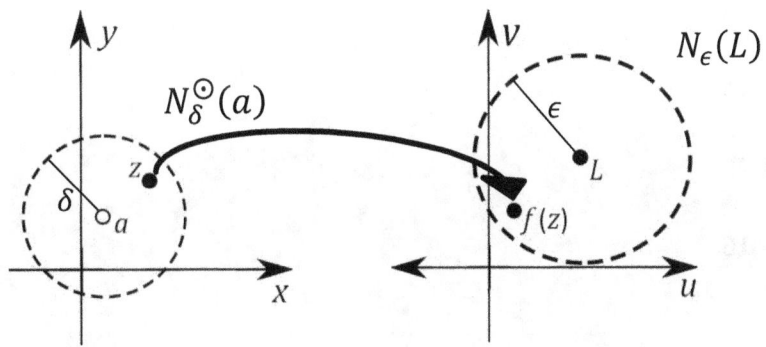

© 2019, All Rights Reserved

Table of Contents

Problem Set 1	7
Problem Set 2	15
Problem Set 3	21
Problem Set 4	33
Problem Set 5	40
Problem Set 6	48
Problem Set 7	56
Problem Set 8	65
Problem Set 9	72
Problem Set 10	79
Problem Set 11	90
Problem Set 12	103
Problem Set 13	109
Problem Set 14	117
Problem Set 15	127
Problem Set 16	142
About the Author	*153*
Books by Dr. Steve Warner	*154*

Problem Set 1

LEVEL 1

1. Determine whether each of the following sentences is an atomic statement, a compound statement, or not a statement at all: (i) I am not going to work today. (ii) What is the meaning of life? (iii) Don't go away mad. (iv) I watched the television show Parks and Recreation. (v) If pigs have wings, then they can fly. (vi) $3 < -5$ or $38 > 37$. (vii) This sentence has five words. (viii) I cannot swim, but I can run fast.

Solutions:

(i) This is a **compound statement**. It has the form $\neg p$, where p is the statement "I am going to work today."

(ii) This is **not a statement**. It is a question.

(iii) This is **not a statement**. It is a command.

(iv) This is an **atomic statement**. Even though the word "and" appears in the statement, here it is part of the name of the show. It is not being used as a connective.

(v) This is a **compound statement**. It has the form $p \to q$, where p is the statement "Pigs have wings," and q is the statement "Pigs can fly."

(vi) This is a **compound statement**. It has the form $p \lor q$, where p is the statement "$3 < -5$" and q is the statement "$38 > 37$."

(vii) This is **not a statement** because it is self-referential. Self-referential sentences can cause problems. For example, observe that the negation of this sentence would be "This sentence does not have five words." The sentence and its negation both appear to be true. That would be a problem. It's a good thing they're not statements!

(viii) This is a **compound statement**. It has the form $\neg p \land q$, where p is the statement "I can swim," and q is the statement "I can run fast." Note that in sentential logic, the word "but" has the same meaning as the word "and." In English, the word "but" is used to introduce contrast with the part of the sentence that has already been mentioned. However, logically it is no different from "and."

2. What is the negation of each of the following statements: (i) The banana is my favorite fruit. (ii) $7 > -3$. (iii) You are not alone. (iv) The function f is differentiable everywhere.

Solutions:

(i) The banana is not my favorite fruit.

(ii) $7 \leq -3$

(iii) You are alone.

(iv) The function f is not differentiable everywhere.

Notes: (1) Technically speaking, the negation of $7 > -3$ (7 is greater than -3) is $7 \not> -3$ (7 is **not** greater than -3). However, this last statement has the same meaning as $7 \leq -3$. In other words, the statements "$7 \not> -3$" and "$7 \leq -3$" are **logically equivalent** (see Note 3 after the solution to Problem 3 below for a precise definition of "logically equivalent."

(2) Similarly, strictly speaking, the negation of "You are not alone" is "It is not the case that you are not alone." This last statement is logically equivalent to the statement "You are alone."

(3) The logical equivalence $p \equiv \neg(\neg p)$ is called the **law of double negation**. We are using this law in the solutions to parts ii and iii above. See Notes 1 and 2 above for an explanation of how this law is being applied. See Example 9.3 for more details on the law of double negation.

LEVEL 2

3. Let p represent the statement "9 is a perfect square," let q represent the statement "Orange is a primary color," and let r represent the statement "A frog is a reptile." Rewrite each of the following symbolic statements in words, and state the truth value of each statement: (i) $p \wedge q$; (ii) $\neg r$; (iii) $p \to r$; (iv) $q \leftrightarrow r$; (v) $\neg p \wedge q$; (vi) $\neg(p \wedge q)$; (vii) $\neg p \vee \neg q$; (viii) $(p \wedge q) \to r$

Solutions: First note that p is true ($9 = 3^2$), q is false (the primary colors are red, yellow and blue; orange is a secondary color), and r is false (a frog is an amphibian, not a reptile).

(i) $p \wedge q$ represents "**9 is a perfect square and orange is a primary color.**" Since q has truth value F, it follows that $p \wedge q$ has truth value **F**.

(ii) $\neg r$ represents "**A frog is not a reptile.**" Since r has truth value F, it follows that $\neg r$ has truth value **T**.

(iii) $p \to r$ represents "**If 9 is a perfect square, then a frog is a reptile.**" Since p has truth value T and r has truth value F, it follows that $p \to r$ has truth value **F**.

(iv) $q \leftrightarrow r$ represents "**Orange is a primary color if and only if a frog is a reptile.**" Since q and r have the same truth value (they both have truth value F), $q \leftrightarrow r$ has truth value **T**.

(v) $\neg p \wedge q$ represents "**9 is not a perfect square and orange is a primary color**" (note that $\neg p \wedge q$ means $(\neg p) \wedge q$). As in (i) above, since q has truth value F, it follows that $\neg p \wedge q$ has truth value **F**.

(vi) $\neg(p \wedge q)$ represents "**It is not the case that 9 is a perfect square and orange is a primary color.**" Since $p \wedge q$ has truth value F (see (i) above), it follows that $\neg(p \wedge q)$ has truth value **T**.

(vii) $\neg p \vee \neg q$ represents "**9 is not a perfect square or orange is not a primary color**" (note that $\neg p \vee \neg q$ means $(\neg p) \vee (\neg q)$). Since $\neg q$ has truth value T (do you see why?), it follows that $\neg p \vee \neg q$ has truth value **T**.

(viii) $(p \wedge q) \to r$ represents "**If 9 is a perfect square and orange is a primary color, then a frog is a reptile.**" Since $p \wedge q$ has truth value F (see (i) above), it follows that $(p \wedge q) \to r$ has truth value **T**.

Notes: (1) In parts 7 and 8 of Example 1.6, we saw an application of one of **De Morgan's laws**, namely that $\neg(p \vee q)$ is logically equivalent to $\neg p \wedge \neg q$. In parts (vi) and (vii) above, we see an application of the other of De Morgan's laws, namely that $\neg(p \wedge q)$ is logically equivalent to $\neg p \vee \neg q$.

(2) Let's draw a full truth table that proves the first De Morgan's law.

p	q	$\neg p$	$\neg q$	$p \vee q$	$\neg(p \vee q)$	$\neg p \wedge \neg q$
T	T	F	F	T	F	F
T	F	F	T	T	F	F
F	T	T	F	T	F	F
F	F	T	T	F	T	T

Observe that the sixth column of the truth table corresponds to $\neg(p \vee q)$, the last (seventh) column corresponds to $\neg p \wedge \neg q$, and both these columns have the same truth values.

(3) When two statements give the same truth values for every assignment of the propositional variables, we say that the statements are **logically equivalent**.

We use the symbol "≡" to indicate logical equivalence. Specifically, if ϕ and ψ are logically equivalent statements, we write $\phi \equiv \psi$.

The truth table above shows that $\neg(p \vee q) \equiv \neg p \wedge \neg q$.

(4) Similarly, the following truth table shows that $\neg(p \wedge q) \equiv \neg p \vee \neg q$.

p	q	$\neg p$	$\neg q$	$p \wedge q$	$\neg(p \wedge q)$	$\neg p \vee \neg q$
T	T	F	F	T	F	F
T	F	F	T	F	T	T
F	T	T	F	F	T	T
F	F	T	T	F	T	T

4. Consider the compound sentence "You can have a cookie or ice cream." In English this would most likely mean that you can have one or the other but not both. The word "or" used here is generally called an "exclusive or" because it excludes the possibility of both. The disjunction is an "inclusive or." Using the symbol $\oplus$ for exclusive or, draw the truth table for this connective.

Solution:

p	q	$p \oplus q$
T	T	F
T	F	T
F	T	T
F	F	F

LEVEL 3

5. Let p, q, and r represent true statements. Compute the truth value of each of the following compound statements: (i) $(p \lor q) \lor r$; (ii) $(p \lor q) \land \neg r$; (iii) $\neg p \to (q \lor r)$; (iv) $\neg(p \leftrightarrow \neg q) \land r$; (v) $\neg[p \land (\neg q \to r)]$; (vi) $\neg[(\neg p \lor \neg q) \leftrightarrow \neg r]$; (vii) $p \to (q \to \neg r)$; (viii) $\neg[\neg p \to (q \to \neg r)]$

Solutions:

(i) $(p \lor q) \lor r \equiv (T \lor T) \lor T \equiv T \lor T \equiv \mathbf{T}$.

(ii) $(p \lor q) \land \neg r \equiv (T \lor T) \land \neg T \equiv T \land F \equiv \mathbf{F}$.

(iii) $\neg p \to (q \lor r) \equiv \neg T \to (T \lor T) \equiv F \to T \equiv \mathbf{T}$.

(iv) $\neg(p \leftrightarrow \neg q) \land r \equiv \neg(T \leftrightarrow \neg T) \land T \equiv \neg(T \leftrightarrow F) \land T \equiv \neg F \land T \equiv T \land T \equiv \mathbf{T}$.

(v) $\neg[p \land (\neg q \to r)] \equiv \neg[T \land (\neg T \to T)] \equiv \neg[T \land (F \to T)] \equiv \neg[T \land T] \equiv \neg T \equiv \mathbf{F}$.

(vi) $\neg[(\neg p \lor \neg q) \leftrightarrow \neg r] \equiv \neg[(\neg T \lor \neg T) \leftrightarrow \neg T] \equiv \neg[(F \lor F) \leftrightarrow F] \equiv \neg[F \leftrightarrow F] \equiv \neg T \equiv \mathbf{F}$.

(vii) $p \to (q \to \neg r) \equiv T \to (T \to \neg T) \equiv T \to (T \to F) \equiv T \to F \equiv \mathbf{F}$.

(viii) $\neg[\neg p \to (q \to \neg r)] \equiv \neg[\neg T \to (T \to \neg T)] \equiv \neg[F \to (T \to F)] \equiv \neg[F \to F] \equiv \neg T \equiv \mathbf{F}$.

Notes: (1) We began each of these problems by replacing the propositional variables p, q, and r by their given truth values (all T). We could save a little time in each case by replacing the negations of each of the propositional variables by F right away. For example. (ii) above would look as follows:

(ii) $(p \lor q) \land \neg r \equiv (T \lor T) \land F \equiv T \land F \equiv \mathbf{F}$.

(2) At each step, we used the truth table of the appropriate connective. For example, in problem (v) to get $\neg[T \land (F \to T)] \equiv \neg[T \land T]$, we used the third row of the truth table for the conditional.

p	q	$p \to q$
T	T	T
T	F	F
F	**T**	**T**
F	F	T

We see from the highlighted row that $F \to T \equiv T$, and therefore $\neg[T \land (\mathbf{F} \to \mathbf{T})] \equiv \neg[T \land \mathbf{T}]$.

Quicker solutions:

(i) $(p \lor q) \lor r \equiv (p \lor q) \lor T \equiv \mathbf{T}$.

(ii) $(p \lor q) \land \neg r \equiv (p \lor q) \land F \equiv \mathbf{F}$.

(iii) $\neg p \to (q \lor r) \equiv F \to (q \lor r) \equiv \mathbf{T}$.

(iv) $\neg(p \leftrightarrow \neg q) \land r \equiv \neg(T \leftrightarrow F) \land T \equiv \neg F \land T \equiv T \land T \equiv \mathbf{T}$.

(v) $\neg[p \land (\neg q \to r)] \equiv \neg[p \land (F \to r)] \equiv \neg[p \land T] \equiv \neg[T \land T] \equiv \neg T \equiv \mathbf{F}$.

(vi) $\neg[(\neg p \vee \neg q) \leftrightarrow \neg r] \equiv \neg[F \leftrightarrow F] \equiv \neg T \equiv \mathbf{F}$.

(vii) $p \rightarrow (q \rightarrow \neg r) \equiv T \rightarrow (T \rightarrow F) \equiv T \rightarrow F \equiv \mathbf{F}$.

(viii) $\neg[\neg p \rightarrow (q \rightarrow \neg r)] \equiv \neg[F \rightarrow (q \rightarrow \neg r)] \equiv \neg T \equiv \mathbf{F}$.

6. Using only the logical connectives $\neg$, $\wedge$, and $\vee$, produce a statement using the propositional variables p and q that has the same truth values as $p \oplus q$ (this is the "exclusive or" defined in Problem 4 above).

Solution: We want to express that p is true or q is true, but p and q are not both true. Expressed in symbols, this is $(\boldsymbol{p} \vee \boldsymbol{q}) \wedge \neg(\boldsymbol{p} \wedge \boldsymbol{q})$.

Note: (1) Let's check that $(p \vee q) \wedge \neg(p \wedge q)$ behaves as desired.

If p and q are both true, then $\neg(p \wedge q) \equiv F$, and so $(p \vee q) \wedge \neg(p \wedge q) \equiv (p \vee q) \wedge F \equiv F$.

If p and q are both false, then $p \vee q \equiv F$, and so $(p \vee q) \wedge \neg(p \wedge q) \equiv F \wedge \neg(p \wedge q) \equiv F$.

Finally, if p and q have opposite truth values, then $p \vee q \equiv T$ and $\neg(p \wedge q) \equiv T$ (because $p \wedge q \equiv F$). Therefore, $(p \vee q) \wedge \neg(p \wedge q) \equiv T \wedge T \equiv T$.

(2) Recall that the word "but" is logically the same as the word "and" (see Problem 1, part (viii)).

(3) Another way to see that $p \oplus q$ has the same truth values as $(p \vee q) \wedge \neg(p \wedge q)$ is to draw the truth tables for each and observe that row by row they have the same truth values. We do this below.

p	q	$p \oplus q$	$p \vee q$	$p \wedge q$	$\neg(p \wedge q)$	$(p \vee q) \wedge \neg(p \wedge q)$
T	T	F	T	T	F	F
T	F	T	T	F	T	T
F	T	T	T	F	T	T
F	F	F	F	F	T	F

Observe that the third column of the truth table corresponds to $p \oplus q$, the last (seventh) column corresponds to $(p \vee q) \wedge \neg(p \wedge q)$, and both these columns have the same truth values.

(4) In this problem, we showed that $p \oplus q \equiv (p \vee q) \wedge \neg(p \wedge q)$ (see Note 3 after Problem 3 above).

LEVEL 4

7. Let p represent a true statement. Decide if this is enough information to determine the truth value of each of the following statements. If so, state that truth value. (i) $p \vee q$; (ii) $p \rightarrow q$; (iii) $\neg p \rightarrow \neg(q \vee \neg r)$; (iv) $\neg(\neg p \wedge q) \leftrightarrow p$; (v) $(p \leftrightarrow q) \leftrightarrow \neg p$; (vi) $\neg[(\neg p \wedge \neg q) \leftrightarrow \neg r]$; (vii) $[(p \wedge \neg p) \rightarrow p] \wedge (p \vee \neg p)$; (viii) $r \rightarrow [\neg q \rightarrow (\neg p \rightarrow \neg r)]$

Solutions:

(i) $(p \vee q) \equiv T \vee q \equiv \mathbf{T}$.

(ii) $p \to q \equiv T \to q$. If $q \equiv T$, we get $T \to T \equiv T$. If $q \equiv F$, we get $T \to F \equiv F$. **There is not enough information**.

(iii) $\neg p \to \neg(q \vee \neg r) \equiv F \to \neg(q \vee \neg r) \equiv T$.

(iv) $\neg(\neg p \wedge q) \leftrightarrow p \equiv \neg(F \wedge q) \leftrightarrow T \equiv \neg F \leftrightarrow T \equiv T \leftrightarrow T \equiv T$.

(v) $(p \leftrightarrow q) \leftrightarrow \neg p \equiv (T \leftrightarrow q) \leftrightarrow F$. If $q \equiv T$, we get $(T \leftrightarrow T) \leftrightarrow F \equiv T \leftrightarrow F \equiv F$. If $q \equiv F$, we get $(T \leftrightarrow F) \leftrightarrow F \equiv F \leftrightarrow F \equiv T$. **There is not enough information**.

(vi) $\neg[(\neg p \wedge \neg q) \leftrightarrow \neg r] \equiv \neg[(F \wedge \neg q) \leftrightarrow \neg r] \equiv \neg(F \leftrightarrow \neg r)$. If $r \equiv T$, we get $\neg T \equiv F$. If $r \equiv F$, we get $\neg F \equiv T$. **There is not enough information**.

(vii) $[(p \wedge \neg p) \to p] \wedge (p \vee \neg p) \equiv [(T \wedge F) \to T] \wedge (T \vee F) \equiv [F \to T] \wedge T \equiv T \wedge T \equiv T$.

(viii) $r \to [\neg q \to (\neg p \to \neg r)] \equiv r \to [\neg q \to (F \to \neg r)] \equiv r \to [\neg q \to T] \equiv r \to T \equiv T$.

8. Assume that the given compound statement is true. Determine the truth value of each propositional variable. (i) $p \wedge q$; (ii) $\neg(p \to q)$; (iii) $p \leftrightarrow [\neg(p \wedge q)]$; (iv) $[p \wedge (q \vee r)] \wedge \neg r$

Solutions:

(i) If $p \equiv F$ or $q \equiv F$, then $p \wedge q \equiv F$. Therefore, $p \equiv T$ and $q \equiv T$.

(ii) Since $\neg(p \to q)$ is true, $p \to q$ is false. Therefore, $p \equiv T$ and $q \equiv F$.

(iii) If $p \equiv F$, then $p \wedge q \equiv F$, and so $p \leftrightarrow [\neg(p \wedge q)] \equiv F \leftrightarrow T \equiv F$. So, $p \equiv T$. It follows that $\neg(p \wedge q) \equiv T$, and so $p \wedge q \equiv F$. Since $p \equiv T$, we must have $q \equiv F$.

(iv) As in (i), we must have $p \wedge (q \vee r) \equiv T$ and $\neg r \equiv T$. So, $p \equiv T$, $q \vee r \equiv T$, and $r \equiv F$. Since $q \vee r \equiv T$ and $r \equiv F$, we must have $q \equiv T$.

LEVEL 5

9. Show that $[p \wedge (q \vee r)] \leftrightarrow [(p \wedge q) \vee (p \wedge r)]$ is always true.

Solution: If $p \equiv F$, then $p \wedge (q \vee r) \equiv F$, $p \wedge q \equiv F$, and $p \wedge r \equiv F$. So, $(p \wedge q) \vee (p \wedge r) \equiv F$. It follows that $[p \wedge (q \vee r)] \leftrightarrow [(p \wedge q) \vee (p \wedge r)] \equiv F \leftrightarrow F \equiv T$.

If $p \equiv T$ and $q \equiv T$, then $p \wedge (q \vee r) \equiv T \wedge T \equiv T$ and $(p \wedge q) \vee (p \wedge r) \equiv T \vee (p \wedge r) \equiv T$. It follows that $[p \wedge (q \vee r)] \leftrightarrow [(p \wedge q) \vee (p \wedge r)] \equiv T \leftrightarrow T \equiv T$.

If $p \equiv T$ and $q \equiv F$, then $p \wedge (q \vee r) \equiv T \wedge r \equiv r$ and $(p \wedge q) \vee (p \wedge r) \equiv F \vee r \equiv r$. It follows that $[p \wedge (q \vee r)] \leftrightarrow [(p \wedge q) \vee (p \wedge r)] \equiv r \leftrightarrow r \equiv T$.

Notes: (1) We can display this reasoning visually as follows:

$$[p \wedge (q \vee r)] \leftrightarrow [(p \wedge q) \vee (p \wedge r)]$$

```
  F F            T   F F   F  F F
  T T T T        T   T T T  T
  T r   F r      T   T F F  r  T r
```

Each row of truth values is placed in the order suggested by the solution above. For example, for the first row, we start by writing F under each p because we are assuming that $p \equiv F$.

$$[p \wedge (q \vee r)] \leftrightarrow [(p \wedge q) \vee (p \wedge r)]$$
$$FFF$$

Next, since the conjunction of F with anything else is F, we write F under each $\wedge$ (there are three that appear).

$$[p \wedge (q \vee r)] \leftrightarrow [(p \wedge q) \vee (p \wedge r)]$$
$$F\ FF\ FF\ F$$

Next, since $F \vee F \equiv F$, we write F under the rightmost $\vee$.

$$[p \wedge (q \vee r)] \leftrightarrow [(p \wedge q) \vee (p \wedge r)]$$
$$F\ FF\ F\ \ F\ F\ F$$

Finally, since $F \leftrightarrow F \equiv T$, we write T under $\leftrightarrow$.

$$[p \wedge (q \vee r)] \leftrightarrow [(p \wedge q) \vee (p \wedge r)]$$
$$F\ FTF\ F\ \ F\ F\ F$$

This is the truth value of the entire statement, and therefore, we are done with the case $p \equiv F$.

The other two rows work the same way.

(2) We could write out the entire truth table for $[p \wedge (q \vee r)] \leftrightarrow [(p \wedge q) \vee (p \wedge r)]$, as was done in the third solution to Example 1.8. This would be an admittedly more tedious way to solve this problem. I leave this solution to the reader.

(3) A statement that has truth value T for all truth assignments of the propositional variables is called a **tautology**. This problem shows us that $[p \wedge (q \vee r)] \leftrightarrow [(p \wedge q) \vee (p \wedge r)]$ is a tautology.

(4) Recall from Note 3 after Problem 3 that two statements are **logically equivalent** if every assignment of the propositional variables leads to the same truth value for both statements. Since $p \leftrightarrow q \equiv T$ if and only if p and q have the same truth value, we see that for statements ϕ and ψ, $\phi \equiv \psi$ if and only if $\phi \leftrightarrow \psi \equiv T$ for every possible combination of truth assignments of the propositional variables appearing in ϕ or ψ if and only if $\phi \leftrightarrow \psi$ is a tautology.

Since $[p \wedge (q \vee r)] \leftrightarrow [(p \wedge q) \vee (p \wedge r)]$ is a tautology, we see that the two statements $p \wedge (q \vee r)$ and $(p \wedge q) \vee (p \wedge r)$ are logically equivalent. This particular equivalence is one of the **distributive laws**. We say that the conjunction is distributive over the disjunction, or $\wedge$ is distributive over $\vee$.

The other distributive law says that $\vee$ is distributive over $\wedge$, so that $p \vee (q \wedge r) \equiv (p \vee q) \wedge (p \vee r)$.

(5) We used two other laws during the third part of the solution: $T \wedge r \equiv r$ and $F \vee r \equiv r$. These are sometimes known as **identity laws**.

(6) See List 9.1 in Lesson 9 for a list of the most common laws.

10. Show that $[[(p \wedge q) \to r] \to s] \to [(p \to r) \to s]$ is always true.

Solution: If $s \equiv T$, then $(p \to r) \to s \equiv T$, and therefore, $[[(p \wedge q) \to r] \to s] \to [(p \to r) \to s] \equiv \mathbf{T}$.

Now, assume $s \equiv F$, and either $p \equiv F$ or $q \equiv F$. Then $p \wedge q \equiv F$, and so $(p \wedge q) \to r \equiv T$. Therefore, $[(p \wedge q) \to r] \to s \equiv F$, and so, $[[(p \wedge q) \to r] \to s] \to [(p \to r) \to s] \equiv \mathbf{T}$.

Finally, assume $s \equiv F$, $p \equiv T$, and $q \equiv T$. Then $p \wedge q \equiv T$, and so, $(p \wedge q) \to r \equiv r$. Therefore, $[(p \wedge q) \to r] \to s \equiv \neg r$. Also, $p \to r \equiv r$, and so $(p \to r) \to s \equiv \neg r$. So, we get $\neg r \to \neg r \equiv \mathbf{T}$.

Notes: (1) The dedicated reader should display this reasoning visually, as was done in Problem 9 above. A full truth table can also be constructed to solve this problem.

(2) This problem says that the statement $[[(p \wedge q) \to r] \to s] \to [(p \to r) \to s]$ is a tautology (see Note 3 after Problem 9).

Problem Set 2

LEVEL 1

1. Determine whether each of the following statements is true or false: (i) $2 \in \{2\}$; (ii) $5 \in \emptyset$; (iii) $\emptyset \in \{1, 2\}$; (iv) $a \in \{b, \{a\}\}$; (v) $\emptyset \subseteq \{1, 2\}$; (vi) $\{\Delta\} \subseteq \{\delta, \Delta\}$; (vii) $\{a, b, c\} \subseteq \{a, b, c\}$; (viii) $\{1, a, \{2, b\}\} \subseteq \{1, a, 2, b\}$

Solutions:

(i) $\{2\}$ has exactly 1 element, namely 2. So, $2 \in \{2\}$ is **true**.

(ii) The empty set has no elements. In particular, $5 \notin \emptyset$. So $5 \in \emptyset$ is **false**.

(iii) $\{1, 2\}$ has 2 elements, namely 1 and 2. Since $\emptyset$ is not one of these, $\emptyset \in \{1, 2\}$ is **false**.

(iv) $\{b, \{a\}\}$ has 2 elements, namely b and $\{a\}$. Since a is not one of these, $a \in \{b, \{a\}\}$ is **false**.

(v) The empty set is a subset of every set. So, $\emptyset \subseteq \{1, 2\}$ is **true**.

(vi) The only element of $\{\Delta\}$ is Δ. Since Δ is also an element of $\{\delta, \Delta\}$, $\{\Delta\} \subseteq \{\delta, \Delta\}$ is **true**.

(vii) Every set is a subset of itself. So, $\{a, b, c\} \subseteq \{a, b, c\}$ is **true**.

(viii) $\{2, b\} \in \{1, a, \{2, b\}\}$, but $\{2, b\} \notin \{1, a, 2, b\}$. So, $\{1, a, \{2, b\}\} \subseteq \{1, a, 2, b\}$ is **false**.

2. Determine the cardinality of each of the following sets: (i) $\{a, b, c, d, e, f\}$; (ii) $\{1, 2, 3, 2, 1\}$; (iii) $\{1, 2, \ldots, 53\}$; (iv) $\{5, 6, 7, \ldots, 2076, 2077\}$

Solutions:

(i) $|\{a, b, c, d, e, f\}| = \mathbf{6}$.

(ii) $\{1, 2, 3, 2, 1\} = \{1, 2, 3\}$. Therefore, $|\{1, 2, 3, 2, 1\}| = |\{1, 2, 3\}| = \mathbf{3}$.

(iii) $|\{1, 2, \ldots, 53\}| = \mathbf{53}$.

(iv) $|\{5, 6, 7, \ldots, 2076, 2077\}| = 2077 - 5 + 1 = \mathbf{2073}$.

Note: For number (iv), we used the fence-post formula (see Notes 3 and 4 after Example 2.6).

3. Let $A = \{a, b, \Delta, \delta\}$ and $B = \{b, c, \delta, \gamma\}$. Determine each of the following: (i) $A \cup B$; (ii) $A \cap B$

Solutions:

(i) $A \cup B = \{a, b, c, \Delta, \delta, \gamma\}$.

(ii) $A \cap B = \{b, \delta\}$.

LEVEL 2

4. Determine whether each of the following statements is true or false: (i) $\emptyset \in \emptyset$; (ii) $\emptyset \in \{\emptyset\}$; (iii) $\{\emptyset\} \in \emptyset$; (iv) $\{\emptyset\} \in \{\emptyset\}$; (v) $\emptyset \subseteq \emptyset$; (vi) $\emptyset \subseteq \{\emptyset\}$; (vii) $\{\emptyset\} \subseteq \emptyset$; (viii) $\{\emptyset\} \subseteq \{\emptyset\}$

Solutions:

(i) The empty set has no elements. So, $x \in \emptyset$ is false for any x. In particular, $\emptyset \in \emptyset$ is **false**.

(ii) The set $\{\emptyset\}$ has exactly 1 element, namely $\emptyset$. So, $\emptyset \in \{\emptyset\}$ is **true**.

(iii) The empty set has no elements. So, $x \in \emptyset$ is false for any x. In particular, $\{\emptyset\} \in \emptyset$ is **false**.

(iv) The set $\{\emptyset\}$ has 1 element, namely $\emptyset$. Since $\{\emptyset\} \neq \emptyset$, $\{\emptyset\} \in \{\emptyset\}$ is **false**.

(v) The empty set is a subset of every set. So, $\emptyset \subseteq X$ is true for any X. In particular, $\emptyset \subseteq \emptyset$ is **true**. (This can also be done by using the fact that every set is a subset of itself.)

(vi) Again, (as in (v)), $\emptyset \subseteq X$ is true for any X. In particular, $\emptyset \subseteq \{\emptyset\}$ is **true**.

(vii) The only subset of $\emptyset$ is $\emptyset$. So, $\{\emptyset\} \subseteq \emptyset$ is **false**.

(viii) Every set is a subset of itself. So, $\{\emptyset\} \subseteq \{\emptyset\}$ is **true**.

5. Determine the cardinality of each of the following sets: (i) $\{\emptyset, \{1, 2, 3\}\}$; (ii) $\{\{\{\emptyset, \{\emptyset\}\}\}\}$; (iii) $\{\{1,2\}, \emptyset, \{\emptyset\}, \{\emptyset, \{\emptyset, 1, 2\}\}\}$; (iv) $\{\emptyset, \{\emptyset\}, \{\{\emptyset\}\}, \{\emptyset, \{\emptyset\}, \{\{\emptyset\}\}\}\}$

Solutions:

(i) The elements of $\{\emptyset, \{1, 2, 3\}\}$ are $\emptyset$ and $\{1, 2, 3\}$. So, we see that $|\{\emptyset, \{1, 2, 3\}\}| = \mathbf{2}$.

(ii) The only element of $\{\{\{\emptyset, \{\emptyset\}\}\}\}$ is $\{\{\emptyset, \{\emptyset\}\}\}$. So, $|\{\{\{\emptyset, \{\emptyset\}\}\}\}| = \mathbf{1}$.

(iii) The elements of $\{\{1,2\}, \emptyset, \{\emptyset\}, \{\emptyset, \{\emptyset, 1, 2\}\}\}$ are $\{1, 2\}$, $\emptyset$, $\{\emptyset\}$, and $\{\emptyset, \{\emptyset, 1, 2\}\}$. So, we see that $|\{\{1,2\}, \emptyset, \{\emptyset\}, \{\emptyset, \{\emptyset, 1, 2\}\}\}| = \mathbf{4}$.

(iv) The elements of $\{\emptyset, \{\emptyset\}, \{\{\emptyset\}\}, \{\emptyset, \{\emptyset\}, \{\{\emptyset\}\}\}\}$ are $\emptyset$, $\{\emptyset\}$, $\{\{\emptyset\}\}$, and $\{\emptyset, \{\emptyset\}, \{\{\emptyset\}\}\}$. So, we see that $|\{\emptyset, \{\emptyset\}, \{\{\emptyset\}\}, \{\emptyset, \{\emptyset\}, \{\{\emptyset\}\}\}\}| = \mathbf{4}$.

6. Let $P = \{\emptyset, \{\emptyset\}\}$ and $Q = \{\{\emptyset\}, \{\emptyset, \{\emptyset\}\}\}$. Determine each of the following: (i) $P \cup Q$; (ii) $P \cap Q$

Solutions:

(i) $P \cup Q = \{\emptyset, \{\emptyset\}, \{\emptyset, \{\emptyset\}\}\}$.

(ii) $P \cap Q = \{\{\emptyset\}\}$.

LEVEL 3

7. How many subsets does $\{a, b, c, d\}$ have? Draw a tree diagram for the subsets of $\{a, b, c, d\}$.

Solution: $|\{a, b, c, d\}| = 4$. Therefore, $\{a, b, c, d\}$ has $2^4 = \mathbf{16}$ subsets. We can also say that the size of the power set of $\{a, b, c, d\}$ is 16, that is, $|\mathcal{P}(\{a, b, c, d\})| = 16$. Here is a tree diagram.

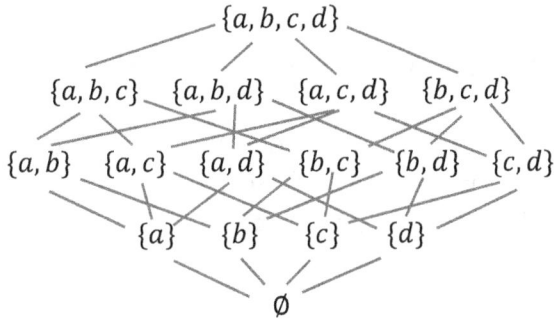

8. A set A is transitive if $\forall x(x \in A \rightarrow x \subseteq A)$ (in words, every element of A is also a subset of A). Determine if each of the following sets is transitive: (i) $\emptyset$; (ii) $\{\emptyset\}$; (iii) $\{\{\emptyset\}\}$; (iv) $\{\emptyset, \{\emptyset\}, \{\{\emptyset\}\}\}$

Solutions:

(i) Since $\emptyset$ has no elements, $\emptyset$ **is transitive**. ($x \in \emptyset \rightarrow x \subseteq \emptyset$ is vacuously true because $x \in \emptyset$ is false).

(ii) The only element of $\{\emptyset\}$ is $\emptyset$, and $\emptyset \subseteq \{\emptyset\}$ (the empty set is a subset of every set). So, $\{\emptyset\}$ **is transitive**.

(iii) $\{\emptyset\} \in \{\{\emptyset\}\}$, but $\{\emptyset\} \not\subseteq \{\{\emptyset\}\}$ because $\emptyset \in \{\emptyset\}$, but $\emptyset \notin \{\{\emptyset\}\}$. So, $\{\{\emptyset\}\}$ **is not transitive**.

(iv) $\{\emptyset, \{\emptyset\}, \{\{\emptyset\}\}\}$ has 3 elements, namely $\emptyset$, $\{\emptyset\}$, and $\{\{\emptyset\}\}$. Let's check each one. $\emptyset$ is a subset of every set. So, $\emptyset \subseteq \{\emptyset, \{\emptyset\}, \{\{\emptyset\}\}\}$. The only element of $\{\emptyset\}$ is $\emptyset$, and $\emptyset \in \{\emptyset, \{\emptyset\}, \{\{\emptyset\}\}\}$. So, $\{\emptyset\} \subseteq \{\emptyset, \{\emptyset\}, \{\{\emptyset\}\}\}$. The only element of $\{\{\emptyset\}\}$ is $\{\emptyset\}$, and $\{\emptyset\} \in \{\emptyset, \{\emptyset\}, \{\{\emptyset\}\}\}$. So, $\{\{\emptyset\}\} \subseteq \{\emptyset, \{\emptyset\}, \{\{\emptyset\}\}\}$. It follows that $\{\emptyset, \{\emptyset\}, \{\{\emptyset\}\}\}$ **is transitive**.

LEVEL 4

9. A relation R is **reflexive** if $\forall x(xRx)$ and **symmetric** if $\forall x \forall y(xRy \rightarrow yRx)$. For example, the relation "=" is reflexive and symmetric because $\forall x(x = x)$ and $\forall x \forall y(x = y \rightarrow y = x)$. Show that $\subseteq$ is reflexive, but $\in$ is not. Then decide if each of $\subseteq$ and $\in$ is symmetric.

Solutions: ($\subseteq$ **is reflexive**) Let A be a set. By Theorem 2.1, A is a subset of itself. So, $A \subseteq A$ is true. Since A was arbitrary, $\forall x(A \subseteq A)$ is true. Therefore, $\subseteq$ is reflexive. □

($\in$ **is not** reflexive) Since the empty set has no elements, $\emptyset \notin \emptyset$. This **counterexample** shows that $\in$ is not reflexive.

($\subseteq$ is **not** symmetric) $\{1\} \subseteq \{1,2\}$, but $\{1,2\} \nsubseteq \{1\}$. This **counterexample** shows that $\subseteq$ is not symmetric.

($\in$ is **not** symmetric) $\emptyset \in \{\emptyset\}$, but $\{\emptyset\} \notin \emptyset$. This **counterexample** shows that $\in$ is not symmetric.

Note: A **conjecture** is an educated guess. In math, conjectures are made all the time based upon evidence from examples (but examples alone cannot be used to prove a conjecture). A logical argument is usually needed to prove a conjecture, whereas a single **counterexample** is used to disprove a conjecture. For example, $\emptyset \notin \emptyset$ is a counterexample to the conjecture "$\in$ is reflexive."

10. Let $A, B, C, D,$ and E be sets such that $A \subseteq B$, $B \subseteq C$, $C \subseteq D$, and $D \subseteq E$. Prove that $A \subseteq E$.

Proof: Suppose that A, B, C, D, and E are sets such that $A \subseteq B$, $B \subseteq C$, $C \subseteq D$, and $D \subseteq E$. Since $A \subseteq B$ and $B \subseteq C$, by Theorem 2.3, we have $A \subseteq C$. Since $A \subseteq C$ and $C \subseteq D$, again by Theorem 2.3, we have $A \subseteq D$. Finally, since $A \subseteq D$ and $D \subseteq E$, once again by Theorem 2.3, we have $A \subseteq E$. □

11. Let A and B be sets. Prove that $A \cap B \subseteq A$.

Proof: Suppose that A and B are sets, and let $x \in A \cap B$. Then $x \in A$ and $x \in B$. In particular, $x \in A$. Since x was an arbitrary element of A, we have shown that every element of $A \cap B$ is an element of A. That is, $\forall x(x \in A \cap B \to x \in A)$ is true. Therefore, $A \cap B \subseteq A$. □

LEVEL 5

12. Let $P(x)$ be the property $x \notin x$. Prove that $\{x | P(x)\}$ cannot be a set.

Solution: Suppose toward contradiction that $A = \{x \,|\, x \notin x\}$ is a set. Then $A \in A$ if and only if $A \notin A$. So, $p \leftrightarrow \neg p$ is true, where p is the statement $A \in A$. However, $p \leftrightarrow \neg p$ is always false. This is a contradiction. So, A is not a set. □

Notes: (1) This is our first **proof by contradiction**. A proof by contradiction works as follows:

1. We assume the negation of what we are trying to prove.
2. We use a logically sound argument to derive a statement which is false.
3. Since the argument is logically sound, the only possible error is our original assumption. Therefore, the negation of our original assumption must be true.

In this problem we are trying to prove that $A = \{x \,|\, x \notin x\}$ **is not** a set. The negation of this statement is that $A = \{x \,|\, x \notin x\}$ **is** a set. We then use only the definition of A to get the false statement $A \in A \leftrightarrow \neg A \in A$. Since the argument was logically valid, our initial assumption must have been incorrect, and therefore A is not a set.

(2) The contradiction that occurs here is known as **Russell's Paradox**. This contradiction shows that we need to be careful about how we define a set. A naïve definition would be that a set is any object that has the form $\{x|P(x)\}$, where $P(x)$ is an arbitrary property (by property, we mean a **first-order property**—this is a property defined using the connectives $\wedge, \vee, \to,$ and $\leftrightarrow$, the quantifiers $\forall$ and $\exists$, and the relations $=$ and $\in$). As we see in this problem, that "definition" of a set leads to a contradiction. Instead, we call $\{x|P(x)\}$ a **class**. Every set is a class, but not every class is a set. A class that is not a set is called a **proper class**. For example, $\{x | x \notin x\}$ is a proper class.

13. Prove that $B \subseteq A$ if and only if $A \cap B = B$.

Proof: Suppose that $B \subseteq A$. By an argument similar to that given in the solution to Problem 11, $A \cap B \subseteq B$. Let $x \in B$. Since $B \subseteq A$, we have $x \in A$. Therefore, $x \in A$ and $x \in B$. So, $x \in A \cap B$. Since x was an arbitrary element of B, we have shown that every element of B is an element of $A \cap B$. That is, $\forall x(x \in B \to x \in A \cap B)$. Therefore, $B \subseteq A \cap B$. Since $A \cap B \subseteq B$ and $B \subseteq A \cap B$, it follows that $A \cap B = B$.

Now, suppose that $A \cap B = B$ and let $x \in B$. Then $x \in A \cap B$. So, $x \in A$ and $x \in B$. In particular, $x \in A$. Since x was an arbitrary element of B, we have shown that every element of B is an element of A. That is, $\forall x(x \in B \to x \in A)$. Therefore, $B \subseteq A$. □

14. Let $A = \{a, b, c, d\}$, $B = \{X \mid X \subseteq A \land d \notin X\}$, and $C = \{X \mid X \subseteq A \land d \in X\}$. Show that there is a natural one-to-one correspondence between the elements of B and the elements of C. Then generalize this result to a set with $n + 1$ elements for $n > 0$.

Solution: We define the one-to-one correspondence as follows: If $Y \in B$, then Y is a subset of A that does not contain d. Let Y_d be the set that contains the same elements as Y, but with d thrown in. Then the correspondence $Y \to Y_d$ is a one-to-one correspondence. We can see this correspondence in the table below.

Elements of B	Elements of C
$\emptyset$	$\{d\}$
$\{a\}$	$\{a, d\}$
$\{b\}$	$\{b, d\}$
$\{c\}$	$\{c, d\}$
$\{a, b\}$	$\{a, b, d\}$
$\{a, c\}$	$\{a, c, d\}$
$\{b, c\}$	$\{b, c, d\}$
$\{a, b, c\}$	$\{a, b, c, d\}$

For the general result, we start with a set A with $n + 1$ elements, and we let d be some element from A. Define B and C the same way as before: $B = \{X \mid X \subseteq A \land d \notin X\}$, and $C = \{X \mid X \subseteq A \land d \in X\}$. Also, as before, if $Y \in B$, then Y is a subset of A that does not contain d. Let Y_d be the set that contains the same elements as Y, but with d thrown in. Then the correspondence $Y \to Y_d$ is a one-to-one correspondence.

Notes: (1) B consists of the subsets of A that do not contain the element d, while C consists of the subsets of A that do contain d.

(2) Observe that in the case where $A = \{a, b, c, d\}$, B and C each have $8 = 2^3$ elements. Also, there is no overlap between B and C (they have no elements in common). So, we have a total of $8 + 8 = 16$ elements. Since there are exactly $2^4 = 16$ subsets of A, we see that we have listed every subset of A.

(3) We could also do the computation in Note 2 as follows: $2^3 + 2^3 = 2 \cdot 2^3 = 2^1 \cdot 2^3 = 2^{1+3} = 2^4$. It's nice to see the computation this way because it mimics the computation we will do in the more general case. In case your algebra skills are not that strong, here is an explanation of each step:

Adding the same thing to itself is equivalent to multiplying that thing by 2. For example, 1 apple plus 1 apple is 2 apples. Similarly, $1x + 1x = 2x$. This could be written more briefly as $x + x = 2x$. Replacing x by 2^3 gives us $2^3 + 2^3 = 2 \cdot 2^3$ (the first equality in the computation above).

Next, by definition, $x^1 = x$. So, $2^1 = 2$. Therefore, we can rewrite $2 \cdot 2^3$ as $2^1 \cdot 2^3$.

Now, 2^3 means to multiply 2 by itself 3 times. So, $2^3 = 2 \cdot 2 \cdot 2$. Thus, $2^1 \cdot 2^3 = 2 \cdot 2 \cdot 2 \cdot 2 = 2^4$. This leads to the rule of exponents which says that if you multiply two expressions with the same base, you can add the exponents. So, $2^1 \cdot 2^3 = 2^{1+3} = 2^4$.

(4) In the more general case, B and C each have 2^n elements. The reason for this is that A has $n + 1$ elements. When we remove the element d from A, the resulting set has n elements, and therefore, 2^n subsets. B consists of precisely the subsets of this new set (A with d removed), and so, B has exactly 2^n elements. The one-to-one correspondence $Y \to Y_d$ shows that C has the same number of elements as B. Therefore, C also has 2^n elements.

(5) In the general case, there is still no overlap between B and C. It follows that the total number of elements when we combine B and C is $2^n + 2^n = 2 \cdot 2^n = 2^1 \cdot 2^n = 2^{1+n} = 2^{n+1}$. See Note 3 above for an explanation as to how all this algebra works.

(6) By a **one-to-one correspondence** between the elements of B and the elements of C, we mean a pairing where we match each element of B with exactly one element of C so that each element of C is matched with exactly one element of B. The table given in the solution above provides a nice example of such a pairing.

(7) In the case where $A = \{a, b, c, d\}$, B consists of all the subsets of $\{a, b, c\}$. In other words, $B = \{X \mid X \subseteq \{a, b, c\}\} = \mathcal{P}(\{a, b, c\})$.

A description of C is a bit more complicated. It consists of the subsets of $\{a, b, c\}$ with d thrown into them. We could write this as $C = \{X \cup \{d\} \mid X \subseteq \{a, b, c\}\}$.

(5) In the general case, we can write $K = A \setminus \{d\}$ (this is the set consisting of all the elements of A, except d). We then have $B = \{X \mid X \subseteq K\} = \mathcal{P}(K)$ and $C = \{X \cup \{d\} \mid X \subseteq K\} = \mathcal{P}(A) \setminus \mathcal{P}(K)$.

(6) The symbol "\" for **set difference** will be defined formally in Lesson 6.

Problem Set 3

LEVEL 1

1. For each of the following multiplication tables defined on the set $S = \{a, b\}$, determine if each of the following is true or false: (i) $\star$ defines a binary operation on S. (ii) $\star$ is commutative in S. (iii) a is an identity with respect to $\star$. (iv) b is an identity with respect to $\star$.

I $\star$	a	b
a	a	a
b	a	a

II $\star$	a	b
a	a	b
b	c	a

III $\star$	a	b
a	a	b
b	b	a

IV $\star$	a	b
a	a	a
b	b	b

Solutions:

(i) For tables I, III, and IV, $\star$ **does** define a binary operation because only a and b appear inside each of these tables. For table II, $\star$ does **not** define a binary operation because an element different from a and b appears in the table (assuming that $c \neq a$ and $c \neq b$).

(ii) For commutativity, since there are just two elements a and b, we need only check if a and b commute ($a \star b = b \star a$). This is very easy to see just by looking at the tables. We simply check if the entries on opposite sides of the main diagonal are the same.

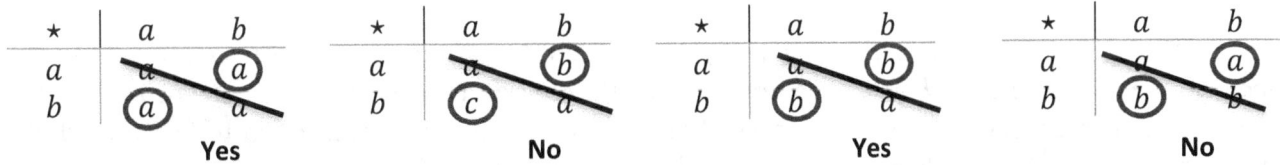

We see that for tables I and III, $\star$ **is** commutative for S, whereas for tables II and IV, $\star$ **is not** commutative for S.

(iii) To see if a is an identity with respect to $\star$, we need to check if $a \star a = a$, $a \star b = b$, and $b \star a = b$. This is also very easy to see just by looking at the tables. We simply check if the row corresponding to a is the same as the "input row," and if the column corresponding to a is the same as the "input column."

$\star$	a	b
a	a	a
b	a	a

No

$\star$	a	b
a	a	b
b	c	a

Maybe

$\star$	a	b
a	a	b
b	b	a

Maybe

$\star$	a	b
a	a	a
b	b	b

No

We see that for tables I and IV, the row corresponding to a is **not** the same as the "input row." So, for I and IV, a is **not** an identity with respect to $\star$.

We still need to check the columns for tables II and III

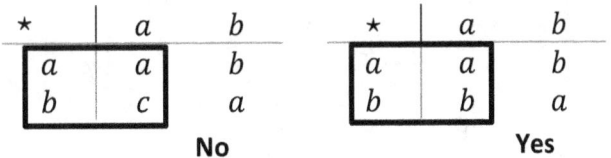

We see that for table II, the column corresponding to a is **not** the same as the "input column." So, for II, a is **not** an identity with respect to $\star$.

For table III, a **is** an identity with respect to $\star$.

(iv) To see if b is an identity with respect to $\star$, we need to check if $a \star b = a$, $b \star a = a$, and $b \star b = b$. Again, this is very easy to see just by looking at the tables. In this case, we see that for each table, the row corresponding to b is **not** the same as the "input row."

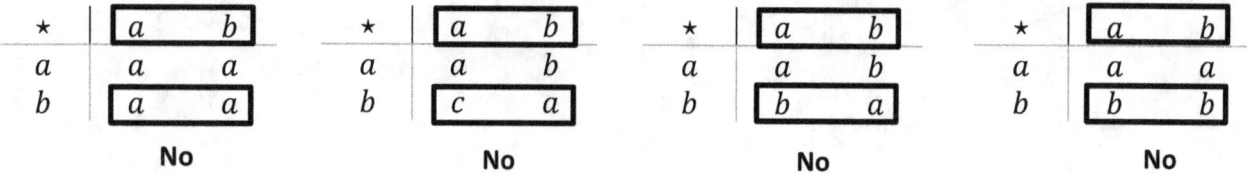

$\star$	a	b
a	a	a
b	a	a

No

$\star$	a	b
a	a	b
b	c	a

No

$\star$	a	b
a	a	b
b	b	a

No

$\star$	a	b
a	a	a
b	b	b

No

So, b is **not** an identity with respect to $\star$ in all four cases.

Notes: (1) Table I defines a semigroup $(S,\star)$. To see that $\star$ is associative in S, just observe that all the outputs are the same. Therefore, there cannot be a counterexample to associativity. For example, $(a \star b) \star b = a \star b = a$ and $a \star (b \star b) = a \star a = a$.

(2) Table I does **not** define a monoid. Parts (iii) and (iv) showed us that there is no identity with respect to $\star$.

(3) Table III defines a commutative group $(S,\star)$ with identity a. a and b are each their own inverses because $a \star a = a$ and $b \star b = a$ (remember that a is the identity). With your current knowledge, associativity can be checked by brute force. There are eight equations that need to be verified. For example, $(a \star a) \star b = a \star b = b$ and $a \star (a \star b) = a \star b = b$. So, $(a \star a) \star b = a \star (a \star b)$. See the solution to Problem 2 below for details.

(4) Table IV defines a semigroup $(S,\star)$ known as the **left zero semigroup**. The name of this semigroup comes from the fact that $a \star a = a$ and $a \star b = a$, so that a is behaving just like 0 behaves when multiplying on the left (0 times anything equals 0). Notice that $b \star a = b \neq a$, so that a does **not** behave like 0 when multiplying on the right. Similar computations show that b also behaves like 0 from the left. The dedicated reader may want to check associativity by brute force, as described in Note 3.

(5) Table IV does **not** define a monoid. Parts (iii) and (iv) showed us that there is no identity with respect to $\star$.

2. Show that there are exactly two monoids on the set $S = \{e, a\}$, where e is the identity. Which of these monoids are groups? Which of these monoids are commutative?

Solution: Let's let e be the identity. Since $e \star x = x \star e = x$ for all x in the monoid, we can easily fill out the first row and the first column of the table.

$\star$	e	a
e	e	a
a	a	$\boxdot$

The entry labeled with ⊡ must be either e or a because we need $\star$ to be a binary operation on S.

Case 1: If we let ⊡ be a, we get the following table.

$\star$	e	a
e	e	a
a	a	a

Associativity holds because any computation of the form $(x \star y) \star z$ or $x \star (y \star z)$ will result in a if any of x, y, or z is a. So, all that is left to check is that $(e \star e) \star e = e \star (e \star e)$. But each side of that equation is equal to e.

So, with this multiplication table, $(S,\star)$ **is** a monoid.

This monoid is **not** a group because a has no inverse. Indeed, $a \star e = a \neq e$ and $a \star a = a \neq e$.

This monoid **is** commutative because $a \star e = a$ and $e \star a = a$.

Case 2: If we let ⊡ be e, we get the following table.

$\star$	e	a
e	e	a
a	a	e

Let's check that associativity holds. There are eight instances to check.

$$(e \star e) \star e = e \star e = e \qquad e \star (e \star e) = e \star e = e$$
$$(e \star e) \star a = e \star a = a \qquad e \star (e \star a) = e \star a = a$$
$$(e \star a) \star e = a \star e = a \qquad e \star (a \star e) = e \star a = a$$
$$(a \star e) \star e = a \star e = a \qquad a \star (e \star e) = a \star e = a$$
$$(e \star a) \star a = a \star a = e \qquad e \star (a \star a) = e \star e = e$$
$$(a \star e) \star a = a \star a = e \qquad a \star (e \star a) = a \star a = e$$
$$(a \star a) \star e = e \star e = e \qquad a \star (a \star e) = a \star a = e$$
$$(a \star a) \star a = e \star a = a \qquad a \star (a \star a) = a \star e = a$$

So, with this multiplication table, $(S,\star)$ **is** a monoid.

Since $e \star e = e$, e is its own inverse. Since $a \star a = e$, a is also its own inverse. Therefore, each element of this monoid is invertible. It follows that this monoid **is** a group.

This monoid **is** commutative because $a \star e = a$ and $e \star a = a$.

3. Let $G = \{e, a, b\}$ and let $(G,\star)$ be a group with identity element e. Draw a multiplication table for $(G,\star)$.

Solution: Since $e \star x = x \star e = x$ for all x in the group, we can easily fill out the first row and the first column of the table.

⋆	e	a	b
e	e	a	b
a	a	⊡	
b	b		

Now, the entry labeled with ⊡ must be either e or b because a is already in that row. If it were e, then the final entry in the row would be b giving two b's in the last column. Therefore, the entry labeled with ⊡ must be b.

⋆	e	a	b
e	e	a	b
a	a	b	
b	b		

Since the same element cannot be repeated in any row or column, the rest of the table is now determined.

⋆	e	a	b
e	e	a	b
a	a	b	e
b	b	e	a

Notes: (1) Why can't the same element appear twice in any row? Well if x appeared twice in the row corresponding to y, that would mean that there are elements z and w with $\mathbf{z \neq w}$ such that $y \star z = x$ and $y \star w = x$. So, $y \star z = y \star w$. We can multiply each side of the equation on the left by y^{-1} (the inverse of y) to get $y^{-1} \star (y \star z) = y^{-1} \star (y \star w)$. By associativity, $(y^{-1} \star y) \star z = (y^{-1} \star y) \star w$. Now, $y^{-1} \star y = e$ by the inverse property. So, we have $e \star z = e \star w$. Finally, since e is an identity, $\mathbf{z = w}$. This contradiction establishes that no element x can appear twice in the same row of a group multiplication table.

A similar argument can be used to show that the same element cannot appear twice in any column.

(2) The argument given in Note 1 used all the group properties (associativity, identity, and inverse). What if we remove one of the properties? For example, what about the multiplication table for a monoid? Can an element appear twice in a row or column? We already know that this can happen—just look at the solution to Problem 2 above.

(3) In Note 1 above, we showed that in the multiplication table for a group, the same element cannot appear as the output more than once in any row or column. We can also show that every element must appear in every row and column. Let's show that the element y must appear in the row corresponding to x. We are looking for an element z such that $x \star z = y$. Well, $z = x^{-1} \star y$ works. Indeed, we have $x \star (x^{-1} \star y) = (x \star x^{-1}) \star y = e \star y = y$.

(4) Using Notes 1 and 3, we see that each element of a group appears exactly once in every row and column of the group's multiplication table.

(5) We have shown that there is essentially just one group of size 3, namely the one given by the table that we produced. Any other group with 3 elements will look exactly like this one, except for possibly the names of the elements. In technical terms, we say that any two groups of order 3 are **isomorphic**.

(6) Observe that in the table we produced, $b = a \star a$. We will generally abbreviate $a \star a$ as a^2. So, another way to draw the table is as follows:

$\star$	e	a	a^2
e	e	a	a^2
a	a	a^2	e
a^2	a^2	e	a

This group is the **cyclic group of order 3**. We call it **cyclic** because the group consists of all powers of the single element a (the elements are a, a^2, and $a^3 = a^0 = e$). The **order** is the number of elements in the group.

> 4. Prove that in any monoid $(M,\star)$, the identity element is unique.

Proof: Let $(M,\star)$ be a monoid, and suppose that e and f are both identity elements in M. Then, we have $f = e \star f = e$. Since we have shown f and e to be equal, there is only one identity element. □

Notes: (1) The word "unique" means that there is only one. In mathematics, we often show that an object is unique by starting with two such objects and then arguing that they must actually be the same. Notice that in the proof above, when we said that e and f are both identity elements, we never insisted that they be *distinct* identity elements. And in fact, the end of the argument shows that they are not distinct.

(2) $e \star f = f$ because e is an identity element and $e \star f = e$ because f is an identity element.

LEVEL 3

> 5. Assume that a group $(G,\star)$ of order 4 exists with $G = \{e, a, b, c\}$, where e is the identity, $a^2 = b$ and $b^2 = e$. Construct the table for the operation of such a group.

Solution: Since $e \star x = x \star e = x$ for all x in the group, we can easily fill out the first row and the first column of the table.

$\star$	e	a	b	c
e	e	a	b	c
a	a			
b	b			
c	c			

We now add in $a \star a = a^2 = b$ and $b \star b = b^2 = e$.

25

★	e	a	b	c
e	e	a	b	c
a	a	b	▫	
b	b		e	
c	c			

Now, the entry labeled with ▫ cannot be a or b because a and b appear in that row. It also cannot be e because e appears in that column. Therefore, the entry labeled with ▫ must be c. It follows that the entry to the right of ▫ must be e, and the entry at the bottom of the column must be a.

★	e	a	b	c
e	e	a	b	c
a	a	b	c	e
b	b	⊙	e	
c	c		a	

Now, the entry labeled with ⊙ cannot be b or e because b and e appear in that row. It also cannot be a because a appears in that column. Therefore, the entry labeled with ⊙ must be c. The rest of the table is then determined.

★	e	a	b	c
e	e	a	b	c
a	a	b	c	e
b	b	c	e	a
c	c	e	a	b

Note: Observe that in the table we produced, $b = a \star a = a^2$ and $c = b \star a = a^2 \star a = a^3$. So, another way to draw the table is as follows:

★	e	a	a^2	a^3
e	e	a	a^2	a^3
a	a	a^2	a^3	e
a^2	a^2	a^3	e	a
a^3	a^3	e	a	a^2

This group is the **cyclic group of order 4**.

6. Prove that in any group $(G, \star)$, each element has a unique inverse.

Proof: Let $a \in G$ and suppose that $b, c \in G$ are both inverses of a. We will show that b and c must be the same. We have $c = c \star e = c \star (a \star b) = (c \star a) \star b = e \star b = b$. □

Notes: (1) $c = c \star e$ because e is an identity element.

(2) $e = a \star b$ because b is an inverse of a. So, $c \star e = c \star (a \star b)$.

(3) $c \star (a \star b) = (c \star a) \star b$ because $\star$ is associative in G.

(4) $c \star a = e$ because c is an inverse of a. So, $(c \star a) \star b = e \star b$.

(5) $e \star b = b$ because e is an identity element.

LEVEL 4

7. Let $(G, \star)$ be a group with $a, b \in G$, and let a^{-1} and b^{-1} be the inverses of a and b, respectively. Prove (i) $(a \star b)^{-1} = b^{-1} \star a^{-1}$; (ii) the inverse of a^{-1} is a.

Proof of (i): Let $a, b \in G$. Then we have

$$(a \star b) \star (b^{-1} \star a^{-1}) = a \star \left(b \star (b^{-1} \star a^{-1})\right) = a \star \left((b \star b^{-1}) \star a^{-1}\right) = a \star (e \star a^{-1}) = a \star a^{-1} = e$$

and

$$(b^{-1} \star a^{-1}) \star (a \star b) = b^{-1} \star \left(a^{-1} \star (a \star b)\right) = b^{-1} \star \left((a^{-1} \star a) \star b\right) = b^{-1} \star (e \star b) = b^{-1} \star b = e.$$

So, $(a \star b)^{-1} = (b^{-1} \star a^{-1})$. □

Notes: (1) For the first and second equalities we used the associativity of $\star$ in G.

(2) For the third equality, we used the inverse property of $\star$ in G.

(3) For the fourth equality, we used the identity property of $\star$ in G.

(4) For the last equality, we again used the inverse property of $\star$ in G.

(5) Since multiplying $a \star b$ on either side by $b^{-1} \star a^{-1}$ results in the identity element e, it follows that $b^{-1} \star a^{-1}$ is the inverse of $a \star b$.

(6) In a group, to verify that an element h is the inverse of an element g, it suffices to show that $g \star h = e$ **or** $h \star g = e$. In other words, we can prove that $g \star h = e \rightarrow h \star g = e$ and we can prove that $h \star g = e \rightarrow g \star h = e$.

For a proof that $g \star h = e \rightarrow h \star g = e$, suppose that $g \star h = e$ and k is the inverse of g. Then $g \star k = k \star g = e$. Since $g \star h = e$ and $g \star k = e$, we have $g \star h = g \star k$. By multiplying by g^{-1} on each side of this equation, and using associativity, the inverse property, and the identity property, we get $h = k$. So, h is in fact the inverse of g.

Proving that $h \star g = e \rightarrow g \star h = e$ is similar. Thus, in the solution above, we need only show one of the sequences of equalities given. The second one follows for free.

Proof of (ii): Let $a \in G$. Since a^{-1} is the inverse of a, we have $a \star a^{-1} = a^{-1} \star a = e$. But this sequence of equations also says that a is the inverse of a^{-1}. □

8. Let $(G, \star)$ be a group such that $a^2 = e$ for all $a \in G$. Prove that $(G, \star)$ is commutative.

Proof: Let $a, b \in G$. Then $(a \star a) \star (b \star b) = a^2 \star b^2 = e \star e = e = (a \star b)^2 = (a \star b) \star (a \star b)$. So, we have $(a \star a) \star (b \star b) = (a \star b) \star (a \star b)$.

We multiply on the left by a^{-1} and on the right by b^{-1} to get

$$a^{-1} \star (a \star a \star b \star b) \star b^{-1} = a^{-1} \star (a \star b \star a \star b) \star b^{-1}$$
$$(a^{-1} \star a) \star a \star b \star (b \star b^{-1}) = (a^{-1} \star a) \star b \star a \star (b \star b^{-1})$$
$$(e \star a) \star (b \star e) = (e \star b) \star (a \star e)$$
$$a \star b = b \star a$$

□

Note: To make the proof less tedious, we have omitted some of the parentheses starting in the fourth line. The associativity of $\star$ allows us to do this. In general, there are two possible meanings for the expression $x \star y \star z$. It could mean $(x \star y) \star z$ or it could mean $x \star (y \star z)$. Since both meanings produce the same result (by associativity), we can simply write $x \star y \star z$ without worrying about the notation being unclear.

We have done this several times in the proof above. For example, we wrote $a \star a \star b \star b$ in the fourth line. This could have multiple meanings, but all those meanings lead to the same result.

9. Prove that $(\mathbb{Q}^*, \cdot)$ is a commutative group.

Proof: (Closure) Let $x, y \in \mathbb{Q}^*$. Then there exist $a, b, c, d \in \mathbb{Z}^*$ such that $x = \frac{a}{b}$ and $y = \frac{c}{d}$. We have $xy = \frac{a}{b} \cdot \frac{c}{d} = \frac{ac}{bd}$. Since $\mathbb{Z}^*$ is closed under multiplication, $ac, bd \in \mathbb{Z}^*$. Therefore, $xy \in \mathbb{Q}^*$.

(Associativity) Let $x, y, z \in \mathbb{Q}^*$. Then there exist $a, b, c, d, e, f \in \mathbb{Z}^*$ such that $x = \frac{a}{b}, y = \frac{c}{d}$, and $z = \frac{e}{f}$. Since multiplication is associative in $\mathbb{Z}^*$, we have

$$(xy)z = \left(\frac{a}{b} \cdot \frac{c}{d}\right)\frac{e}{f} = \left(\frac{ac}{bd}\right)\frac{e}{f} = \frac{(ac)e}{(bd)f} = \frac{a(ce)}{b(df)} = \frac{a}{b}\left(\frac{ce}{df}\right) = \frac{a}{b}\left(\frac{c}{d} \cdot \frac{e}{f}\right) = x(yz).$$

(Identity) Let $\bar{1} = \frac{1}{1}$. We show that $\bar{1}$ is an identity for $(\mathbb{Q}^*, \cdot)$. Let $x \in \mathbb{Q}^*$. Then there exist $a, b \in \mathbb{Z}^*$ such that $x = \frac{a}{b}$. Since 1 is an identity for $\mathbb{Z}^*$, we have

$$x \cdot \bar{1} = \frac{a}{b} \cdot \frac{1}{1} = \frac{a \cdot 1}{b \cdot 1} = \frac{a}{b} = x \text{ and } \bar{1}x = \frac{1}{1} \cdot \frac{a}{b} = \frac{1a}{1b} = \frac{a}{b} = x.$$

(Inverse) Let $x \in \mathbb{Q}^*$. Then there exist $a, b \in \mathbb{Z}^*$ such that $x = \frac{a}{b}$. Let $y = \frac{b}{a}$. Then $y \in \mathbb{Q}^*$ (note that $a \neq 0$). Since multiplication is commutative in $\mathbb{Z}^*$, we have

$$xy = \frac{a}{b} \cdot \frac{b}{a} = \frac{ab}{ba} = \frac{ab}{ab} = \frac{1}{1} = \bar{1}.$$

So, y is the multiplicative inverse of x.

(Commutativity) Let $x, y \in \mathbb{Q}^*$. Then there exist $a, b, c, d \in \mathbb{Z}^*$ such that $x = \frac{a}{b}$ and $y = \frac{c}{d}$. Since multiplication is commutative in $\mathbb{Z}^*$, we have

$$xy = \frac{a}{b} \cdot \frac{c}{d} = \frac{ac}{bd} = \frac{ca}{db} = \frac{c}{d} \cdot \frac{a}{b} = yx.$$

So, $(\mathbb{Q}^*, \cdot)$ is a commutative group.

Important note: There is one more issue here. It's not obvious that the definition of multiplication is even well-defined. Suppose that $\frac{a}{b} = \frac{a'}{b'}$ and $\frac{c}{d} = \frac{c'}{d'}$. We need to check that $\frac{a}{b} \cdot \frac{c}{d} = \frac{a'}{b'} \cdot \frac{c'}{d'}$, or equivalently, $\frac{ac}{bd} = \frac{a'c'}{b'd'}$.

Since $\frac{a}{b} = \frac{a'}{b'}$, we have $ab' = ba'$. Since $\frac{c}{d} = \frac{c'}{d'}$, we have $cd' = dc'$. Now, since $ab' = ba'$, $cd' = dc'$, and multiplication is commutative and associative in $\mathbb{Z}$, we have

$$(ac)(b'd') = (ab')(cd') = (ba')(dc') = (bd)(a'c')$$

Therefore, $\frac{ac}{bd} = \frac{a'c'}{b'd'}$, as desired. □

LEVEL 5

10. Prove that there are exactly two groups of order 4, up to renaming the elements.

Solution: Let $G = \{e, a, b, c\}$. We will run through the possible cases.

Case 1: Suppose that $a^2 = e$.

Since $e \star x = x \star e = x$ for all x in the group, we can easily fill out the first row, the first column, and one more entry of the table.

⋆	e	a	b	c
e	e	a	b	c
a	a	e	⊡	
b	b	⊡		
c	c			

Each of the entries labeled with ⊡ cannot be a, e, or b (Why?), and so they must be c. So, we get the following:

⋆	e	a	b	c
e	e	a	b	c
a	a	e	c	b
b	b	c		
c	c	b		

Now, if $b^2 = e$, the rest of the table is determined:

⋆	e	a	b	c
e	e	a	b	c
a	a	e	c	b
b	b	c	e	a
c	c	b	a	e

This table gives a group $(G, \star)$ called the **Klein four group**.

If $b^2 = a$, the rest of the table is also determined:

$\star$	e	a	b	c
e	e	a	b	c
a	a	e	c	b
b	b	c	a	e
c	c	b	e	a

This table gives a group $(G,\star)$ called the **Cyclic group of order 4**.

Observe that we cannot have $b^2 = b$ or $b^2 = c$ because b and c already appear in the row (and column) corresponding to b.

$\star$	e	a	b	c
e	e	a	b	c
a	a	e	c	b
b	b	c	~~b,c~~	
c	c	b		

Case 2: Suppose that $a^2 \neq e$.

If $b^2 = e$ or $c^2 = e$, then by renaming elements, we get the same groups in Case 1. So, we may assume that $a^2 \neq e$, $b^2 \neq e$, and $c^2 \neq e$.

So, a, a^2, and a^3 are distinct elements. If $a^3 \neq e$, then $a^4 = e$, and so, $(a^2)^2 = e$. But a^2 must be equal to either b or c. So, $b^2 = e$ or $c^2 = e$, contrary to our assumption.

It follows that $a^3 = e$. So, $a \star a^2 = e$. Therefore, a and a^2 are inverses of each other. If $a^2 = b$, then c must be its own inverse. So, $c^2 = c \star c = e$, contrary to our assumption. Similarly, if $a^2 = c$, then b must be its own inverse. So, $b^2 = b \star b = e$, contrary to our assumption.

It follows that there are exactly 2 groups of order 4, up to renaming the elements. These 2 groups are the **Klein four group** and the **Cyclic group of order 4**. □

11. Show that $(\mathbb{Q}, +)$ is a commutative group.

Proof: (Closure) Let $x, y \in \mathbb{Q}$. Then there exist $a, c \in \mathbb{Z}$ and $b, d \in \mathbb{Z}^*$ such that $x = \frac{a}{b}$ and $y = \frac{c}{d}$. We have $x + y = \frac{a}{b} + \frac{c}{d} = \frac{ad+bc}{bd}$. Since $\mathbb{Z}$ is closed under multiplication, $ad \in \mathbb{Z}$ and $bc \in \mathbb{Z}$. Since $\mathbb{Z}$ is closed under addition, $ad + bc \in \mathbb{Z}$. Since $\mathbb{Z}^*$ is closed under multiplication, $bd \in \mathbb{Z}^*$. Therefore, $x + y \in \mathbb{Q}$.

(Associativity) Let $x, y, z \in \mathbb{Q}$. Then there exist $a, c, e \in \mathbb{Z}$ and $b, d, f \in \mathbb{Z}^*$ such that $x = \frac{a}{b}$, $y = \frac{c}{d}$, and $z = \frac{e}{f}$. Since multiplication and addition are associative in $\mathbb{Z}$, multiplication is (both left and right) distributive over addition in $\mathbb{Z}$ (see Note 1 below), and multiplication is associative in $\mathbb{Z}^*$, we have

$$(x+y) + z = \left(\frac{a}{b} + \frac{c}{d}\right) + \frac{e}{f} = \frac{ad+bc}{bd} + \frac{e}{f} = \frac{(ad+bc)f + (bd)e}{(bd)f} = \frac{((ad)f + (bc)f) + (bd)e}{(bd)f}$$

$$= \frac{a(df) + (b(cf) + b(de))}{b(df)} = \frac{a(df) + b(cf+de)}{b(df)} = \frac{a}{b} + \frac{cf+de}{df} = \frac{a}{b} + \left(\frac{c}{d} + \frac{e}{f}\right) = x + (y+z).$$

(Identity) Let $\bar{0} = \frac{0}{1}$. We show that $\bar{0}$ is an identity for $(\mathbb{Q}, +)$. Let $x \in \mathbb{Q}$. Then there exist $a \in \mathbb{Z}$ and $b \in \mathbb{Z}^*$ such that $x = \frac{a}{b}$. Since 0 is an identity for $\mathbb{Z}$, and $0 \cdot x = x \cdot 0 = 0$ for all $x \in \mathbb{Z}$, we have

$$x + \bar{0} = \frac{a}{b} + \frac{0}{1} = \frac{a \cdot 1 + b \cdot 0}{b \cdot 1} = \frac{a + 0}{b} = \frac{a}{b} = x \text{ and } \bar{0} + x = \frac{0}{1} + \frac{a}{b} = \frac{0b + 1a}{1b} = \frac{0 + a}{b} = \frac{a}{b} = x.$$

(Inverse) Let $x \in \mathbb{Q}$. Then there exist $a \in \mathbb{Z}$ and $b \in \mathbb{Z}^*$ such that $x = \frac{a}{b}$. Let $y = \frac{-1a}{b}$. Since $\mathbb{Z}$ is closed under multiplication, $-1a \in \mathbb{Z}$. So, $y \in \mathbb{Q}$. Since multiplication is associative and commutative in $\mathbb{Z}$ and $(-1)n = -n$ for all $n \in \mathbb{Z}$, we have

$$x + y = \frac{a}{b} + \frac{-1a}{b} = \frac{ab + b(-1a)}{b \cdot b} = \frac{ab + (-1a)b}{b^2} = \frac{ab + (-1)(ab)}{b^2} = \frac{ab - ab}{b^2} = \frac{0}{b^2} = \bar{0}$$

$$y + x = \frac{-1a}{b} + \frac{a}{b} = \frac{(-1a)b + ba}{b \cdot b} = \frac{-1(ab) + ab}{b^2} = \frac{-ab + ab}{b^2} = \frac{0}{b^2} = \bar{0}$$

So, y is the additive inverse of x.

(Commutativity) Let $x, y \in \mathbb{Q}$. Then there exist $a, c \in \mathbb{Z}$ and $b, d \in \mathbb{Z}^*$ such that $x = \frac{a}{b}$ and $y = \frac{c}{d}$. Since multiplication and addition are commutative in $\mathbb{Z}$, and multiplication is commutative in $\mathbb{Z}^*$, we have

$$x + y = \frac{a}{b} + \frac{c}{d} = \frac{ad + bc}{bd} = \frac{bc + ad}{db} = \frac{cb + da}{db} = \frac{c}{d} + \frac{a}{b} = y + x.$$

So, $(\mathbb{Q}, +)$ is a commutative group. □

Notes: (1) Multiplication is **distributive** over addition in $\mathbb{Z}$. That is, for all $x, y, z \in \mathbb{Z}$, we have

$$x \cdot (y + z) = x \cdot y + x \cdot z \quad \text{and} \quad (y + z) \cdot x = y \cdot x + z \cdot x$$

The first equation is called the **left distributive property** and the second equation is called the **right distributive property**. Since multiplication is commutative in $\mathbb{Z}$, right distributivity follows from left distributivity (and vice versa), and we can simply call either of the two properties the **distributive property**.

For example, if we let $x = 2$, $y = 3$, and $z = 4$, we have

$$2(3 + 4) = 2 \cdot 7 = 14 \text{ and } 2 \cdot 3 + 2 \cdot 4 = 6 + 8 = 14.$$

The picture to the right gives a physical representation of the distributive property for this example. Note that the area of the light grey rectangle is $2 \cdot 3$, the area of the dark grey rectangle is $2 \cdot 4$, and the area of the whole rectangle is $2(3 + 4)$.

When verifying associativity above, we used right distributivity for the fourth equality and left distributivity for the sixth equality. Distributivity will be discussed in more detail in Lesson 4.

(2) As we did for multiplication in Problem 9 above, we need to check that the definition of addition is well-defined. Suppose that $\frac{a}{b} = \frac{a'}{b'}$ and $\frac{c}{d} = \frac{c'}{d'}$. We need to check that $\frac{a}{b} + \frac{c}{d} = \frac{a'}{b'} + \frac{c'}{d'}$, or equivalently, $\frac{ad+bc}{bd} = \frac{a'd'+b'c'}{b'd'}$.

Since $\frac{a}{b} = \frac{a'}{b'}$, we have $ab' = ba'$. Since $\frac{c}{d} = \frac{c'}{d'}$, we have $cd' = dc'$. Now, since $ab' = ba'$, $cd' = dc'$, multiplication is commutative and associative in $\mathbb{Z}$, and multiplication is distributive over addition in $\mathbb{Z}$, we have

$$(ad + bc)(b'd') = adb'd' + bcb'd' = ab'dd' + cd'bb' = ba'dd' + dc'bb'$$
$$= bda'd' + bdb'c' = (bd)(a'd' + b'c').$$

Therefore, $\frac{ad+bc}{bd} = \frac{a'd'+b'c'}{b'd'}$, as desired. □

> 12. Let $S = \{a, b\}$, where $a \neq b$. How many binary operations are there on S? How many semigroups are there of the form $(S, \star)$, up to renaming the elements?

Solution: The number of binary operations is $2^4 = \mathbf{16}$. Let's draw all possible multiplication tables for $(S, \star)$, where $\star: S \times S \to S$ is a binary operation.

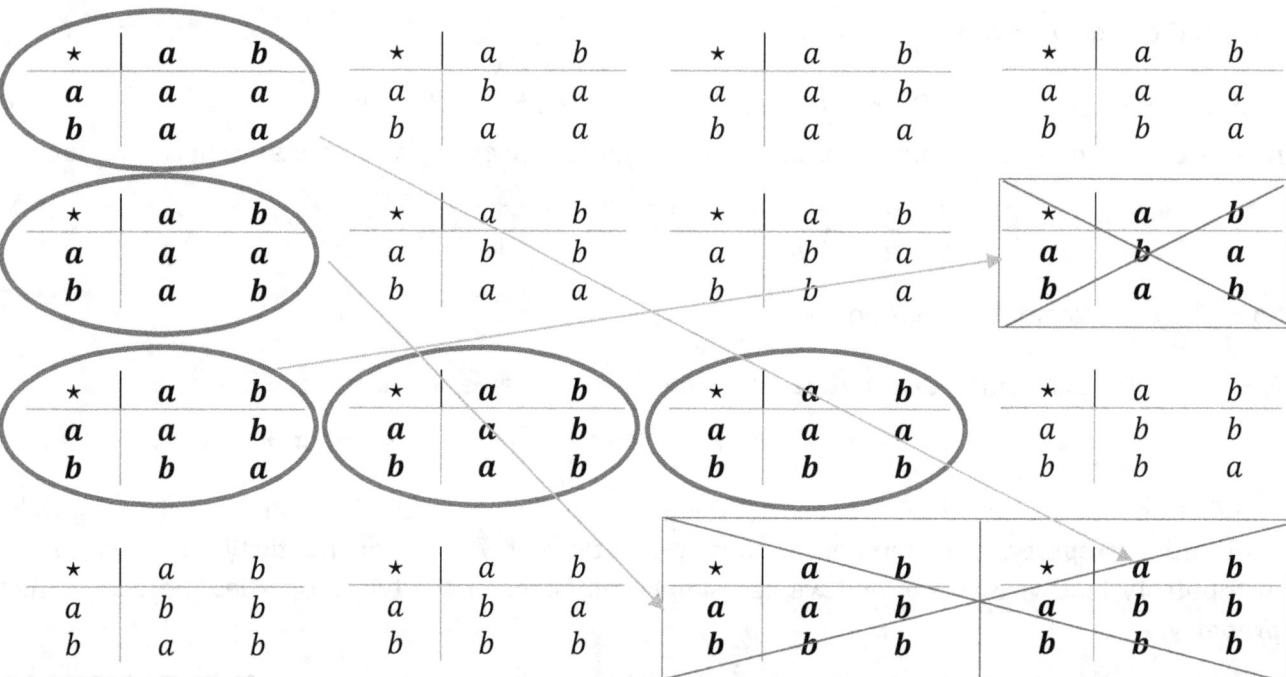

Of the 16 binary operations, 8 give rise to semigroups. However, 3 of these are essentially the same as 3 of the others. The 5 circled multiplication tables represent the 5 semigroups of order 2. The 3 tables in rectangles that are crossed out also represent semigroups. However, if you interchange the roles of a and b you'll see that they are the same as 3 of the others with the names changed (arrows are present to indicate the tables that are essentially the same as these). The other 8 tables represent operations that are not associative (the reader should find a counterexample to associativity for each of these). I leave it to the reader to verify that the 5 circled multiplication tables represent semigroups.

Note: A **magma** is a pair $(M, \star)$, where M is a set and $\star$ is a binary operation on M (and no other conditions). In the solution above we showed that there are 16 magmas of the form $(\{a, b\}, \star)$, and of these, 8 are semigroups. However, there are only 5 semigroups up to renaming the elements. Of the 16 magmas, there are only 10 up to renaming the elements. See if you can find the duplicates.

Problem Set 4

LEVEL 1

1. The addition and multiplication tables below are defined on the set $S = \{0, 1\}$. Show that $(S, +, \cdot)$ does **not** define a ring.

+	0	1
0	0	1
1	1	0

$\cdot$	0	1
0	1	0
1	0	1

Solution: We have $0(1 + 1) = 0 \cdot 0 = 1$ and $0 \cdot 1 + 0 \cdot 1 = 0 + 0 = 0$. So, $0(1+1) \neq 0 \cdot 1 + 0 \cdot 1$. Therefore, multiplication is **not** distributive over addition in S, and so, $(S, +, \cdot)$ does not define a ring.

Notes: (1) Both multiplication tables given are the same, except that we interchanged the roles of 0 and 1 (in technical terms, $(S, +)$ and $(S, \cdot)$ are **isomorphic**).

Both tables represent the unique table for a group with 2 elements. See Problem 2 from Lesson 3 for details.

(2) Since $(S, +)$ is a commutative group and $(S, \cdot)$ is a monoid (in fact, it's a commutative group), we know that the only possible way $(S, +, \cdot)$ can fail to be a ring is for distributivity to fail.

2. Let $S = \{0, 1\}$ and define addition (+) and multiplication ($\cdot$) so that $(S, +, \cdot)$ is a ring. Assume that 0 is the additive identity in S and 1 is the multiplicative identity in S. Draw the tables for addition and multiplication and verify that with these tables, $(S, +, \cdot)$ is a ring.

Solution: Since $(S, +)$ is a commutative group, by Problem 2 in Lesson 3, the addition table must be the following.

+	0	1
0	0	1
1	1	0

Since $(S, \cdot)$ is a monoid and 1 is the multiplicative identity, again by Problem 2 in Lesson 3, the multiplication table must be one of the following.

$\cdot$	0	1
0	1	0
1	0	1

$\cdot$	0	1
0	0	0
1	0	1

However, we showed in Problem 1 that if we use the table on the left, then $(S, +, \cdot)$ will **not** define a ring.

So, the addition and multiplication tables must be as follows:

33

+	0	1
0	0	1
1	1	0

·	0	1
0	0	0
1	0	1

Since we already know that $(S, +)$ is a commutative group and $(S, \cdot)$ is a monoid, all we need to verify is that distributivity holds. Since $\cdot$ is commutative for S (by Problem 2 in Lesson 3), it suffices to verify left distributivity. We will do this by brute force. There are eight instances to check.

$$0(0+0) = 0 \cdot 0 = 0 \qquad 0 \cdot 0 + 0 \cdot 0 = 0 + 0 = 0$$
$$0(0+1) = 0 \cdot 1 = 0 \qquad 0 \cdot 0 + 0 \cdot 1 = 0 + 0 = 0$$
$$0(1+0) = 0 \cdot 1 = 0 \qquad 0 \cdot 1 + 0 \cdot 0 = 0 + 0 = 0$$
$$0(1+1) = 0 \cdot 0 = 0 \qquad 0 \cdot 1 + 0 \cdot 1 = 0 + 0 = 0$$
$$1(0+0) = 1 \cdot 0 = 0 \qquad 1 \cdot 0 + 1 \cdot 0 = 0 + 0 = 0$$
$$1(0+1) = 1 \cdot 1 = 1 \qquad 1 \cdot 0 + 1 \cdot 1 = 0 + 1 = 1$$
$$1(1+0) = 1 \cdot 1 = 1 \qquad 1 \cdot 1 + 1 \cdot 0 = 1 + 0 = 1$$
$$1(1+1) = 1 \cdot 0 = 0 \qquad 1 \cdot 1 + 1 \cdot 1 = 1 + 1 = 0$$

So, we see that left distributivity holds, and therefore $(S, +, \cdot)$ is a ring. □

LEVEL 2

3. Use the Principle of Mathematical Induction to prove the following: (i) $2^n > n$ for all natural numbers $n \geq 1$. (ii) $0 + 1 + 2 + \cdots + n = \frac{n(n+1)}{2}$ for all natural numbers. (iii) $n! > 2^n$ for all natural numbers $n \geq 4$ (where $n! = 1 \cdot 2 \cdots n$ for all natural numbers $n \geq 1$). (iv) $2^n \geq n^2$ for all natural numbers $n \geq 4$.

Proofs:

(i) **Base Case** ($k = 1$): $2^1 = 2 > 1$.

Inductive Step: Let $k \in \mathbb{N}$ with $k \geq 1$ and assume that $2^k > k$. Then we have

$$2^{k+1} = 2^k \cdot 2^1 = 2^k \cdot 2 > k \cdot 2 = 2k = k + k \geq k + 1.$$

Therefore, $2^{k+1} > k + 1$.

By the Principle of Mathematical Induction, $2^n > n$ for all natural numbers $n \geq 1$. □

(ii) **Base Case** ($k = 0$): $0 = \frac{0(0+1)}{2}$.

Inductive Step: Let $k \in \mathbb{N}$ and assume that $0 + 1 + 2 + \cdots + k = \frac{k(k+1)}{2}$. Then we have

$$0 + 1 + 2 + \cdots + k + (k+1) = \frac{k(k+1)}{2} + (k+1) = (k+1)\left(\frac{k}{2} + 1\right) = (k+1)\left(\frac{k}{2} + \frac{2}{2}\right)$$

$$= (k+1)\left(\frac{k+2}{2}\right) = \frac{(k+1)(k+2)}{2} = \frac{(k+1)((k+1)+1)}{2}$$

By the Principle of Mathematical Induction, $0 + 1 + 2 + \cdots + n = \frac{n(n+1)}{2}$ for all natural numbers n. □

(iii) **Base Case** ($k = 4$): $4! = 1 \cdot 2 \cdot 3 \cdot 4 = 24 > 16 = 2^4$.

Inductive Step: Let $k \in \mathbb{N}$ with $k \geq 4$ and assume that $k! > 2^k$. Then we have

$$(k+1)! = (k+1)k! > (k+1)2^k \geq (4+1) \cdot 2^k = 5 \cdot 2^k \geq 2 \cdot 2^k = 2^1 \cdot 2^k = 2^{1+k} = 2^{k+1}.$$

Therefore, $(k+1)! > 2^{k+1}$.

By the Principle of Mathematical Induction, $n! > 2^n$ for all natural numbers $n \geq 4$. □

(iv) **Base Case** ($k = 4$): $2^4 = 16 = 4^2$. So, $2^4 \geq 4^2$.

Inductive Step: Let $k \in \mathbb{N}$ with $k \geq 4$ and assume that $2^k \geq k^2$. Then we have

$$2^{k+1} = 2^k \cdot 2^1 \geq k^2 \cdot 2 = 2k^2 = k^2 + k^2.$$

By Example 4.6, $k^2 > 2k + 1$. So, we have $2^{k+1} > k^2 + 2k + 1 = (k+1)^2$.

Therefore, $2^{k+1} \geq (k+1)^2$.

By the Principle of Mathematical Induction, $2^n \geq n^2$ for all $n \in \mathbb{N}$ with $n \geq 4$. □

Note: Let's take one last look at number (iv). $2^0 = 1 \geq 0 = 0^2$. So, the statement in (iv) is true for $k = 0$. Also, $2^1 = 2 \geq 1 = 1^2$ and $2^2 = 4 = 2^2$. So, the statement is true for $k = 1$ and $k = 2$. However, $2^3 = 8$ and $3^2 = 9$. So, the statement is false for $k = 3$. It follows that $2^n \geq n^2$ for all natural numbers n except $n = 3$.

4. Show that the sum of three integers that are each divisible by 5 is divisible by 5.

Proof: Let m, n, and q be integers that are divisible by 5. Then there are integers j, k, and r such that $m = 5j$, $n = 5k$, and $q = 5r$. So, $m + n + q = 5j + 5k + 5r = 5(j + k) + 5r = 5(j + k + r)$ because multiplication is distributive over addition in $\mathbb{Z}$. Since $\mathbb{Z}$ is closed under addition, we have $j + k + r \in \mathbb{Z}$. Therefore, $m + n + q$ is divisible by 5. □

Note: At this point, as in Problem 8 from Lesson 3 and Example 4.6, we are being more relaxed in our use of associativity. The expression $m + n + q$ makes sense here because addition is associative in $\mathbb{Z}$. In general, $m + n + q$ could mean $(m + n) + q$ or $m + (n + q)$. By associativity, both expressions are equal, and we can leave the parentheses out without causing confusion.

LEVEL 3

5. Prove that if $a, b, c \in \mathbb{Z}$ with $a|b$ and $b|c$, then $a|c$.

Proof: Let $a, b, c \in \mathbb{Z}$ with $a|b$ and $b|c$. Since $a|b$, there is $j \in \mathbb{Z}$ such that $b = aj$. Since $b|c$, there is $k \in \mathbb{Z}$ such that $c = bk$. It follows that $c = bk = (aj)k = a(jk)$ because multiplication is associative in $\mathbb{Z}$. Since $j, k \in \mathbb{Z}$ and $\mathbb{Z}$ is closed under multiplication, $jk \in \mathbb{Z}$. Therefore, $a|c$. □

6. Prove that $n^3 - n$ is divisible by 3 for all natural numbers n.

Proof by Mathematical induction:

Base Case ($k = 0$): $0^3 - 0 = 0 = 3 \cdot 0$. So, $0^3 - 0$ is divisible by 3.

Inductive Step: Let $k \in \mathbb{N}$ and assume that $k^3 - k$ is divisible by 3. Then $k^3 - k = 3b$ for some integer b. Now,

$$(k+1)^3 - (k+1) = (k+1)[(k+1)^2 - 1] = (k+1)[(k+1)(k+1) - 1]$$
$$= (k+1)(k^2 + 2k + 1 - 1) = (k+1)(k^2 + 2k) = k^3 + 2k^2 + k^2 + 2k = k^3 + 3k^2 + 2k$$
$$= k^3 - k + k + 3k^2 + 2k = (k^3 - k) + 3k^2 + 3k = 3b + 3(k^2 + k) = 3(b + k^2 + k).$$

Here we used the fact that $(\mathbb{Z}, +, \cdot)$ is a ring. Since $\mathbb{Z}$ is closed under addition and multiplication, $b + k^2 + k \in \mathbb{Z}$. Therefore, $(k+1)^3 - (k+1)$ is divisible by 3.

By the Principle of Mathematical Induction, $n^3 - n$ is divisible by 3 for all $n \in \mathbb{N}$. □

Notes: (1) Okay...we cheated a little here. Instead of writing out every algebraic step and mentioning every property of the natural numbers we used at each of these steps, we skipped over some of the messy algebra and at the end of it all simply mentioned that all this is okay because $(\mathbb{Z}, +, \cdot)$ is a ring.

For example, we replaced $(k+1)(k+1)$ by $k^2 + 2k + 1$. You may remember this "rule" as FOIL (first, inner, outer, last) from your high school classes. We have not yet verified that FOILing is a legal operation in the set of natural numbers. Let's check the details:

$$(k+1)(k+1) = (k+1) \cdot k + (k+1) \cdot 1 = k \cdot k + k + k + 1 = k^2 + 2k + 1$$

For the first equality, we used left distributivity of multiplication over addition, and for the second equality, we used right distributivity of multiplication over addition, together with the multiplicative identity property and associativity of addition (we've omitted parentheses when adding several terms).

(2) It's a worthwhile exercise to find all the other places in the proof where details were excluded and to fill in those details.

(3) Notice our use of SACT (see Note 7 after Example 4.5) in the beginning of the last line of the sequence of equations. We needed $k^3 - k$ to appear, but the $-k$ was nowhere to be found. So, we simply threw it in, and then repaired the damage by adding k right after it.

LEVEL 4

7. Prove that if $a, b, c, d, e \in \mathbb{Z}$ with $a|b$ and $a|c$, then $a|(db + ec)$.

Proof: Let $a, b, c, d, e \in \mathbb{Z}$ with $a|b$ and $a|c$. Since $a|b$, there is $j \in \mathbb{Z}$ such that $b = aj$. Since $a|c$, there is $k \in \mathbb{Z}$ such that $c = ak$. Since $(\mathbb{Z}, +, \cdot)$ is a ring, it follows that

$$db + ec = d(aj) + e(ak) = (da)j + (ea)k = (ad)j + (ae)k = a(dj) + a(ek) = a(dj + ek).$$

Since $\mathbb{Z}$ is closed under multiplication, $dj \in \mathbb{Z}$ and $ek \in \mathbb{Z}$. Since $\mathbb{Z}$ is closed under addition, $dj + ek \in \mathbb{Z}$. So, $a|(db + ec)$. □

Notes: (1) As in Problem 6, we skipped over mentioning every property of the integers we used at each step, and simply mentioned that $(\mathbb{Z}, +, \cdot)$ is a ring. The dedicated reader may want to fill in the details.

(2) The expression $db + ec$ is called a **linear combination** of b and c. Linear combinations come up often in advanced mathematics and we will see them more in later lessons.

8. Prove that $3^n - 1$ is even for all natural numbers n.

Proof by Mathematical induction:

Base Case ($k = 0$): $3^0 - 1 = 1 - 1 = 0 = 2 \cdot 0$. So, $3^0 - 1$ is even.

Inductive Step: Let $k \in \mathbb{N}$ and assume that $3^k - 1$ is even. Then $3^k - 1 = 2b$ for some integer b. Now,

$$3^{k+1} - 1 = 3^k \cdot 3^1 - 1 = 3^k \cdot 3 - 1 = 3^k \cdot 3 - 3^k + 3^k - 1 = 3^k(3-1) + (3^k - 1)$$
$$= 3^k \cdot 2 + 2b = 2 \cdot 3^k + 2b = 2(3^k + b).$$

Here we used the fact that $(\mathbb{Z}, +, \cdot)$ is a ring. Since $\mathbb{Z}$ is closed under multiplication, $3^k \in \mathbb{Z}$. Since $\mathbb{Z}$ is closed under addition, $3^k + b \in \mathbb{Z}$. Therefore, $3^{k+1} - 1$ is even.

By the Principle of Mathematical Induction, $3^n - 1$ is even for all $n \in \mathbb{N}$. □

Notes: (1) As in Problem 6, we skipped over mentioning every property of the natural numbers we used at each step, and simply mentioned that $(\mathbb{Z}, +, \cdot)$ is a ring. The dedicated student may want to fill in the details.

(2) Notice our use of SACT (see Note 7 after Example 4.5) in the middle of the first line of the sequence of equations. We needed $3^k - 1$ to appear, so we added 3^k, and then subtracted 3^k to the left of it.

9. Show that Theorem 4.3 (the Principle of Mathematical Induction) is equivalent to the following statement:

   ($\star$) Let $P(n)$ be a statement and suppose that (i) $P(0)$ is true and (ii) for all $k \in \mathbb{N}$, $P(k) \to P(k+1)$. Then $P(n)$ is true for all $n \in \mathbb{N}$.

Proof: Recall that Theorem 4.3 says the following: Let S be a set of natural numbers such that (i) $0 \in S$ and (ii) for all $k \in \mathbb{N}$, $k \in S \to k+1 \in S$. Then $S = \mathbb{N}$.

Suppose that Theorem 4.3 is true, and let $P(n)$ be a statement such that $P(0)$ is true, and for all $k \in \mathbb{N}, P(k) \to P(k+1)$. Define $S = \{n \mid (P(n)\}$. Since $P(0)$ is true, $0 \in S$. If $k \in S$, then $P(k)$ is true. So, $P(k+1)$ is true, and therefore, $k+1 \in S$. By Theorem 4.3, $S = \mathbb{N}$. So, $P(n)$ is true for all $n \in \mathbb{N}$.

Now, suppose that ($\star$) holds, and let S be a set of natural numbers such that $0 \in S$, and for all $k \in \mathbb{N}$, $k \in S \to k+1 \in S$. Let $P(n)$ be the statement $n \in S$. Since $0 \in S$, $P(0)$ is true. If $P(k)$ is true, then $k \in S$. So, $k+1 \in S$, and therefore, $P(k+1)$ is true. By ($\star$), $P(n)$ is true for all n. So, for all $n \in \mathbb{N}$, we have $n \in S$. In other words, $\mathbb{N} \subseteq S$. Since we were given $S \subseteq \mathbb{N}$, we have $S = \mathbb{N}$. □

Recall: If A and B are sets, then one way to prove that A and B are equal is to show that each one is a subset of the other.

In the beginning of the third paragraph, we let S be a set of natural numbers. In other words, we are assuming that $S \subseteq \mathbb{N}$. We then argue that we also have $\mathbb{N} \subseteq S$. It then follows that $S = \mathbb{N}$.

See the Technical note after Theorem 2.5 (in Lesson 2) for more details.

LEVEL 5

10. The Principle of Strong Induction is the following statement:

$(\star\star)$ Let $P(n)$ be a statement and suppose that (i) $P(0)$ is true and (ii) for all $k \in \mathbb{N}$, $\forall j \leq k\, (P(j)) \to P(k+1)$. Then $P(n)$ is true for all $n \in \mathbb{N}$.

Use the Principle of Mathematical Induction to prove the Principle of Strong Induction.

Proof: Assume that $(\star)$ from Problem 9 above is true.

Let $P(n)$ be a statement such that $P(0)$ is true, and for all $k \in \mathbb{N}$, $\forall j \leq k\, (P(j)) \to P(k+1)$. Let $Q(n)$ be the statement $\forall j \leq n\, (P(j))$.

Base case: $Q(0) \equiv \forall j \leq 0 (P(j)) \equiv P(0)$. Since $P(0)$ is true and $Q(0) \equiv P(0)$, $Q(0)$ is also true.

Inductive step: Suppose that $Q(k)$ is true. Then $\forall j \leq k\, (P(j))$ is true. Therefore, $P(k+1)$ is true. So $Q(k) \wedge P(k+1)$ is true. But notice that

$$Q(k+1) \equiv \forall j \leq k+1 (P(j)) \equiv \forall j \leq k(P(j)) \wedge P(k+1) \equiv Q(k) \wedge P(k+1).$$

So, $Q(k+1)$ is true.

By the Principle of Mathematical Induction (($\star$) from Problem 9), $Q(n)$ is true for all $n \in \mathbb{N}$. This implies that $P(n)$ is true for all $n \in \mathbb{N}$. $\square$

11. Show that $(\mathbb{Q}, +, \cdot)$ is a field.

Proof: By Problem 11 in Lesson 3, $(\mathbb{Q}, +)$ is a commutative group. By Problem 9 in Lesson 3, $(\mathbb{Q}^*, \cdot)$ is a commutative group. So, all that's left to show is that multiplication is distributive over addition in $\mathbb{Q}$.

(Distributivity) Let $x, y, z \in \mathbb{Q}$. Then there exist $a, c, e \in \mathbb{Z}$ and $b, d, f \in \mathbb{Z}^*$ such that $x = \frac{a}{b}$, $y = \frac{c}{d}$, and $z = \frac{e}{f}$. Let's start with left distributivity.

$$x(y+z) = \frac{a}{b}\left(\frac{c}{d} + \frac{e}{f}\right) = \frac{a}{b}\left(\frac{cf+de}{df}\right) = \frac{a(cf+de)}{b(df)}$$

$$xy + xz = \frac{a}{b} \cdot \frac{c}{d} + \frac{a}{b} \cdot \frac{e}{f} = \frac{ac}{bd} + \frac{ae}{bf} = \frac{(ac)(bf) + (bd)(ae)}{(bd)(bf)}$$

We need to verify that $\frac{(ac)(bf)+(bd)(ae)}{(bd)(bf)} = \frac{a(cf+de)}{b(df)}$.

Since $\mathbb{Z}$ is a ring, $(ac)(bf) + (bd)(ae) = bacf + bade = ba(cf + de)$ (see Note 1 below).

Since multiplication is associative and commutative in $\mathbb{Z}^*$, we have

$$(bd)(bf) = b(d(bf)) = b((db)f) = b((bd)f) = b(b(df)).$$

So, $\frac{(ac)(bf)+(bd)(ae)}{(bd)(bf)} = \frac{ba(cf+de)}{b(b(df))} = \frac{a(cf+de)}{b(df)}$.

For right distributivity, we can use left distributivity together with the commutativity of multiplication in $\mathbb{Q}$.

$$(y + z)x = x(y + z) = xy + xz = yx + zx \qquad \square$$

Notes: (1) We skipped many steps when verifying $(ac)(bf) + (bd)(ae) = ba(cf + de)$. The dedicated reader may want to verify this equality carefully, making sure to use only the fact that $\mathbb{Z}$ is a ring, and noting which ring property is being used at each step.

(2) In the very last step of the proof, we cancelled one b in the numerator of the fraction with b in the denominator of the fraction. In general, if $j \in \mathbb{Z}$ and $m, k \in \mathbb{Z}^*$, then $\frac{mj}{mk} = \frac{j}{k}$. To verify that this is true, simply observe that since $\mathbb{Z}$ is a ring, we have $(mj)k = m(jk) = m(kj) = (mk)j$. (Remember from part 4 of Example 3.6 in Lesson 3 that we identify rational numbers $\frac{a}{b}$ and $\frac{c}{d}$ whenever $ad = bc$).

> 12. Use the Principle of Mathematical Induction to prove that for every $n \in \mathbb{N}$, if S is a set with $|S| = n$, then S has 2^n subsets. (Hint: Use Problem 14 from Lesson 2.)

Proof: Base Case ($k = 0$): Let S be a set with $|S| = 0$. Then $S = \emptyset$, and the empty set has exactly 1 subset, namely itself. So, the number of subsets of S is $1 = 2^0$.

Inductive Step: Assume that for any set S with $|S| = k$, S has 2^k subsets.

Now, let A be a set with $|A| = k + 1$, let d be any element from A, and let $S = A \setminus \{d\}$ (S is the set consisting of all elements of A except d). $|S| = k$, and so, by the inductive hypothesis, S has 2^k subsets. Let $B = \{X \mid X \subseteq A \land d \notin X\}$ and $C = \{X \mid X \subseteq A \land d \in X\}$. B is precisely the set of subsets of S, and so $|B| = 2^k$. By Problem 14 from Lesson 2, $|B| = |C|$ and therefore, $|C| = 2^k$. Also, B and C have no elements in common and every subset of A is in either B or C. So, the number of subsets of A is equal to $|B| + |C| = 2^k + 2^k = 2 \cdot 2^k = 2^1 \cdot 2^k = 2^{1+k} = 2^{k+1}$.

By the Principle of Mathematical Induction, given any $n \in \mathbb{N}$, if S is a set with $|S| = n$, then S has 2^n subsets. $\qquad \square$

Notes: (1) Recall from Lesson 2 that $|S| = n$ means that the set S has n elements.

(2) Also, recall from Lesson 2 that if S is a set, then the **power set** of S is the set of subsets of S.

$$\mathcal{P}(S) = \{X \mid X \subseteq S\}$$

In this problem, we proved that a set with n elements has a power set with 2^n elements. Symbolically, we have

$$|S| = n \rightarrow |\mathcal{P}(S)| = 2^n.$$

Problem Set 5

LEVEL 1

1. The addition and multiplication tables below are defined on the set $S = \{0, 1, 2\}$. Show that $(S, +, \cdot)$ does **not** define a field.

+	0	1	2
0	0	1	2
1	1	2	0
2	2	0	1

·	0	1	2
0	0	0	0
1	0	1	2
2	0	2	2

Solution: We have $2 \cdot 0 = 0$, $2 \cdot 1 = 2$, and $2 \cdot 2 = 2$. So, 2 has no multiplicative inverse, and therefore, $(S, +, \cdot)$ does **not** define a field.

Note: It's not difficult to check that $(S, +)$ is a group with identity 0 and $(S, \cdot)$ is a monoid with identity 1. However, $(S, +, \cdot)$ is not a ring, as distributivity fails. Here is a counterexample:

$$2(1 + 1) = 2 \cdot 2 = 2 \qquad 2 \cdot 1 + 2 \cdot 1 = 2 + 2 = 1$$

We could have also used this computation to verify that $(S, +, \cdot)$ is not a field.

2. Let $F = \{0, 1\}$, where $0 \neq 1$. Show that there is exactly one field $(F, +, \cdot)$, where 0 is the additive identity and 1 is the multiplicative identity.

Solution: Suppose that $(F, +, \cdot)$ is a field. Since $(F, +)$ is a commutative group, by Problem 2 in Lesson 3, the addition table must be the following.

+	0	1
0	0	1
1	1	0

Since $(F^*, \cdot)$ is a monoid and 1 is the multiplicative identity, we must have $1 \cdot 1 = 1$.

Now, if $0 \cdot 0 = 1$, then we have $1 = 0 \cdot 0 = 0(0 + 0) = 0 \cdot 0 + 0 \cdot 0 = 1 + 1 = 0$, a contradiction. So, $0 \cdot 0 = 0$.

If $0 \cdot 1 = 1$, then we have $1 = 0 \cdot 1 = (0 + 0) \cdot 1 = 0 \cdot 1 + 0 \cdot 1 = 1 + 1 = 0$, a contradiction. So, $0 \cdot 1 = 0$.

Finally, if $1 \cdot 0 = 1$, then we have $1 = 1 \cdot 0 = 1(0 + 0) = 1 \cdot 0 + 1 \cdot 0 = 1 + 1 = 0$, a contradiction. So, $1 \cdot 0 = 0$.

It follows that the addition and multiplication tables must be as follows:

+	0	1
0	0	1
1	1	0

·	0	1
0	0	0
1	0	1

We already verified in Problem 2 from Lesson 4 that $(F, +, \cdot)$ is a ring. Since $1 \cdot 1 = 1$, the multiplicative inverse property holds, and it follows that $(F, +, \cdot)$ is a field.

LEVEL 2

3. Let $(F, +, \cdot)$ be a field. Prove each of the following: (i) If $a, b \in F$ with $a + b = b$, then $a = 0$; (ii) If $a \in F$, $b \in F^*$, and $ab = b$, then $a = 1$; (iii) If $a \in F$, then $a \cdot 0 = 0$; (iv) If $a \in F^*$, $b \in F$, and $ab = 1$, then $b = \frac{1}{a}$; (v) If $a, b \in F$ and $ab = 0$, then $a = 0$ or $b = 0$; (vi) If $a \in F$, then $-a = -1a$; (vii) $(-1)(-1) = 1$

Proofs:

(i) Let $a, b \in F$ with $a + b = b$. Then we have
$$a = a + 0 = a + (b + (-b)) = (a + b) + (-b) = b + (-b) = 0.\qquad\square$$

(ii) Let $a \in F$, $b \in F^*$, and $ab = b$. Then we have
$$a = a \cdot 1 = a(bb^{-1}) = (ab)b^{-1} = bb^{-1} = 1.\qquad\square$$

(iii) Let $a \in F$. Then $a \cdot 0 + a = a \cdot 0 + a \cdot 1 = a(0 + 1) = a \cdot 1 = a$. By (i), $a \cdot 0 = 0$. $\square$

(iv) Let $a \in F^*$, $b \in F$, and $ab = 1$. Then $b = 1b = (a^{-1}a)b = a^{-1}(ab) = a^{-1} \cdot 1 = a^{-1} = \frac{1}{a}$. $\square$

(v) Let $a, b \in F$ and $ab = 0$. Assume that $a \neq 0$. Then $b = 1b = (a^{-1}a)b = a^{-1}(ab) = a^{-1} \cdot 0$. By (iii), $a^{-1} \cdot 0 = 0$. So, $b = 0$. $\square$

(vi) Let $a \in F$. Then $-1a + a = a(-1) + a \cdot 1 = a(-1 + 1) = a \cdot 0 = 0$ (by (iii)). So, $-1a$ is the additive inverse of a. Thus, $-1a = -a$. $\square$

(vii) $(-1)(-1) + (-1) = (-1)(-1) + (-1) \cdot 1 = (-1)(-1 + 1) = (-1)(0) = 0$ (by (iii)). So, we see that $(-1)(-1)$ is the additive inverse of -1. Therefore, $(-1)(-1) = -(-1)$. $\square$

Notes: (1) The dedicated reader should provide justification for each equality that appears above. For example, a justification of the sequence of equalities in (i) above could be notated as follows:

$$\underset{\text{IdA}}{a = a + 0} \underset{\text{InA}}{= a + (b + (-b))} \underset{\text{AsA}}{= (a + b) + (-b)} \underset{\text{G}}{= b + (-b)} \underset{\text{InA}}{= 0.}$$

IdA is an abbreviation for the identity property under addition.

InA is an abbreviation for the inverse property under addition.

AsA is an abbreviation for associativity under addition.

Finally, G stands for given (we are given that $a + b = b$).

(2) What if we require only that $(F, +, \cdot)$ be a ring instead of a field. Which of the above results are still true?

4. Let $(F, +, \cdot)$ be a field with $\mathbb{N} \subseteq F$. Prove that $\mathbb{Q} \subseteq F$.

Proof: Let $n \in \mathbb{Z}$. If $n \in \mathbb{N}$, then $n \in F$ because $\mathbb{N} \subseteq F$. If $n \notin \mathbb{N}$, then $-n \in \mathbb{N}$. So, $-n \in F$. Since F is a field, we have $n = -(-n) \in F$. For each $n \in \mathbb{Z}^*$, $\frac{1}{n} = n^{-1} \in F$ because $n \in F$ and the inverse property holds in F. Now, let $\frac{m}{n} \in \mathbb{Q}$. Then $m \in \mathbb{Z}$ and $n \in \mathbb{Z}^*$. Since $\mathbb{Z} \subseteq F$, $m \in F$. Since $n \in \mathbb{Z}^*$, we have $\frac{1}{n} \in F$. Therefore, $\frac{m}{n} = \frac{m \cdot 1}{1 \cdot n} = \frac{m}{1} \cdot \frac{1}{n} = m\left(\frac{1}{n}\right) \in F$ because F is closed under multiplication. Since $\frac{m}{n}$ was an arbitrary element of $\mathbb{Q}$, we see that $\mathbb{Q} \subseteq F$. □

LEVEL 3

5. Let $(F, \leq)$ be an ordered field. Prove each of the following: (i) If $a, b \in F$, exactly one of the following holds: $a < b$, $a = b$, or $a > b$; (ii) If $a, b \in F$, $a \leq b$, and $b \leq a$, then $a = b$; (iii) If $a, b, c \in F$, $a < b$, and $b < c$, then $a < c$; (iv) If $a, b, c \in F$, $a \leq b$, and $b \leq c$, then $a \leq c$; (v) If $a, b \in F^+$ and $a > b$, then $\frac{1}{a} < \frac{1}{b}$; (vi) If $a, b \in F$, then $a > b$ if and only if $-a < -b$; (vii) If $a, b \in F$, then $a \geq b$ if and only if $-a \leq -b$.

Proofs:

(i) Let $a, b \in F$. Since F is a field, $b - a = b + (-a) \in F$. By Order Property (3), exactly one of the following holds: $b - a > 0$, $b - a = 0$, or $-(b - a) > 0$.

We have that $b - a > 0$ is equivalent to $b > a$ or $a < b$ (by definition). Also, $b - a = 0$ is equivalent to $b + (-a) = 0$ or $b = -(-a) = a$. Finally, we have that $-(b - a)$ is equivalent to $-1(b + (-a)) = -1b + (-1)(-a) = -1b + (-1)(-1)a = -1b + 1a = a - b$. So, we see that $-(b - a) > 0$ is equivalent to $a - b > 0$, or $a > b$. □

(ii) Let $a, b \in F$, $a \leq b$, and $b \leq a$. Since $a \leq b$, we have $a < b$ or $a = b$. If $a = b$, we are done. So, assume $a < b$. Since $b \leq a$, we have $b < a$ or $a = b$. By (i), we cannot have $a < b$ and $b < a$. So, we must have $a = b$. □

(iii) Let $a, b, c \in F$, $a < b$, and $b < c$. Then $b - a > 0$ and $c - b > 0$. By Order Property (1), we have $c - a = (c - b) + (b - a) > 0$. So, $c > a$, or equivalently, $a < c$. □

(iv) Let $a, b, c \in F$, $a \leq b$, and $b \leq c$. Since $a \leq b$, we have $a < b$ or $a = b$. First, suppose that $a < b$. Since $b \leq c$, we have $b < c$ or $b = c$. If $b < c$, then by (iii), $a < c$, and so, $a \leq c$. If $b = c$, Then $a < b$ and $b = c$ imples $a < c$ (by substituting b for c), and therefore, $a \leq c$. Next, suppose that $a = b$. If $b < c$, then $a < c$ (by substituting b for a), and thus, $a \leq c$. If $b = c$, then $a = c$ (again by substituting b for a). □

(v) Let $a, b \in F^+$ and $a > b$. Then $a - b > 0$. So, $\frac{1}{b} - \frac{1}{a} = \frac{1}{ab}(a - b)$. Since $a, b \in F^+$, $ab \in F^+$ by Order Property (2). So, $\frac{1}{ab} \in F^+$ by Theorem 5.4. Since $\frac{1}{ab} > 0$ and $a - b > 0$, by Order Property (2), we have $\frac{1}{b} - \frac{1}{a} = \frac{1}{ab}(a - b) > 0$. So, $\frac{1}{b} > \frac{1}{a}$, or equivalently, $\frac{1}{a} < \frac{1}{b}$. □

(vi) Let $a, b \in F$. Then $a > b$ if and only if $a - b > 0$ if and only if $-(a - b) < 0$ if and only if $-1(a + (-b)) < 0$ if and only if $-1a - 1(-b) < 0$ if and only if $-a - (-b) < 0$ if and only if $-a < -b$.

□

(vii) Let $a, b \in F$. Then $a \geq b$ if and only if $a - b \geq 0$ if and only if $a - b > 0$ or $a - b = 0$ if and only if $a > b$ or $a = b$ if and only if $-a < -b$ or $-a = -b$ if and only if $-a - (-b) < 0$ or $-a - (-b) = 0$ if and only if $-a - (-b) \leq 0$. If and only if $-a \leq -b$. □

6. Let $(F, +, \cdot)$ be a field. Show that $(F, \cdot)$ is a commutative monoid.

Proof: Let $(F, +, \cdot)$ be a field. Then $\cdot$ is a binary operation on F and $(F^*, \cdot)$ is a commutative group.

We first show that if $a \in F$, then $0a = 0$. To see this, observe that

$$0a + a = 0a + 1a = (0 + 1)a = 1a = a.$$

By Problem 3, part (i), $0a = 0$.

Let $x, y \in F$. If $x, y \in F^*$, then $xy = yx$. If $x = 0$, then $xy = 0y = 0$ by the previous result, and $yx = y \cdot 0 = 0$ by Problem 3 part (iii) above. If $y = 0$, then $xy = x \cdot 0 = 0$ by Problem 3, part (iii) above, and $yx = 0x = 0$ by the previous result. In all cases, we have $xy = yx$.

Next, let $x, y, z \in F$. If $x, y, z \in F^*$, then $(xy)z = x(yz)$. If $x = 0$, then by the previous result, we have $(xy)z = (0y)z = 0z = 0$ and $x(yz) = 0(yz) = 0$. If $y = 0$, by Problem 3, part (iii) and the previous result, we have $(xy)z = (x \cdot 0)z = 0z = 0$ and $x(yz) = x(0z) = x \cdot 0 = 0$. If $z = 0$, we have $(xy)z = (xy) \cdot 0 = 0$ and $x(yz) = x(y \cdot 0) = x \cdot 0 = 0$. In all cases, we have $(xy)z = x(yz)$.

Let $x \in F$. If $x \in F^*$, then $1x = x \cdot 1 = x$. If $x = 0$, then by Problem 3, part (iii), $1x = 1 \cdot 0 = 0$ and by the previous result, $x \cdot 1 = 0 \cdot 1 = 0$. In all cases, we have $1x = x \cdot 1 = x$.

Therefore, $(F, \cdot)$ is a commutative monoid. □

LEVEL 4

7. Prove that there is no smallest positive real number.

Proof: Let $x \in \mathbb{R}^+$ and let $y = \frac{1}{2}x$. By Theorem 5.4, $\frac{1}{2} > 0$. So, by Order Property (2), $y > 0$.

Now, $x - y = x - \frac{1}{2}x = 1x - \frac{1}{2}x = \left(1 - \frac{1}{2}\right)x = \left(\frac{2}{2} - \frac{1}{2}\right)x = \frac{1}{2}x > 0$. So, $x > y$. It follows that y is a positive real number smaller than x. Since x was an arbitrary positive real number, there is no smallest positive real number. □

8. Let a be a nonnegative real number. Prove that $a = 0$ if and only if a is less than every positive real number. (Note: a nonnegative means that a is positive or zero.)

Proof: Let a be a nonnegative real number.

First suppose that $a = 0$. Let ϵ be a positive real number, so that $\epsilon > 0$. Then by direct substitution, $\epsilon > a$, or equivalently $a < \epsilon$. Since ϵ was an arbitrary positive real number, we have shown that a is less than every positive real number.

Now, suppose that a is less than every positive real number. Assume towards contradiction that $a \neq 0$. Then $a > 0$ (because a is nonnegative). Let $\epsilon = \frac{1}{2}a$. By the same reasoning used in Problem 7 above, we have that ϵ is a positive real number with $a > \epsilon$. This contradicts our assumption that a is less than every positive real number. □

Note: There are three methods for proving a statement of the form $p \to q$:

(i) **Direct proof:** *In a direct proof, we assume p, and deduce q.*

Most of the proofs we have done up to this point have been direct proofs. For example, above we proved the statement "If $a = 0$, then a is less than every positive number" using a direct proof. Notice how the proof starts with "Suppose that $a = 0$," and ends with "a is less than every positive real number."

(ii) **Proof by contradiction:** *In a proof by contradiction, we assume the opposite of what we want to prove, and derive a contradiction.*

The opposite of the statement $p \to q$ is $\neg(p \to q)$, which is **logically equivalent** to the statement $p \wedge \neg q$. This means that all four possible assignments of truth values for the propositional variables p and q lead to the same resulting truth value. For example, if we let p be true and q false, then we have $\neg(p \to q) \equiv \neg(T \to F) \equiv \neg F \equiv T$, and similarly, we have $p \wedge \neg q \equiv T \wedge \neg F \equiv T \wedge T \equiv T$. The dedicated reader should check the other three truth assignments as well.

So, to summarize, to prove a statement of the form $p \to q$ by contradiction, we assume $p \wedge \neg q$, and derive a contradiction.

We have done a few proofs by contradiction so far. For example, above we proved the statement "If a is less than every positive real number, then $a = 0$" by contradiction. Notice how the proof starts with "Suppose that a is less than every positive real number **and** $a \neq 0$."

(iii) **Proof by contrapositive:** The **contrapositive** of the conditional statement $p \to q$ is $\neg q \to \neg p$. The contrapositive of a conditional statement is logically equivalent to the original conditional statement (Check this!).

In a proof by contrapositive, we assume $\neg q$, and deduce $\neg p$.

Some proofs by contradiction can be modified slightly to become proofs by contrapositive. When this can be done, most mathematicians would prefer to use contrapositive over contradiction.

In this problem, a small modification in our proof by contradiction will turn it into a proof by contrapositive. Here is the modified version:

"Now, suppose that $a \neq 0$. Then $a > 0$. Let $\epsilon = \frac{1}{2}a$. By the same reasoning used in Problem 7 above, we have that ϵ is a positive real number with $a > \epsilon$. So, a is **not** less than every positive real number."

Notice how this version the proof starts with "$a \neq 0$" and ends with "a is not less than every positive real number."

9. Prove that every rational number can be written in the form $\frac{m}{n}$, where $m \in \mathbb{Z}$, $n \in \mathbb{Z}^*$, and at least one of m or n is **not** even.

Proof: Let x be a rational number. Then there are $a \in \mathbb{Z}$ and $b \in \mathbb{Z}^*$ such that $x = \frac{a}{b}$. Let j be the largest integer such that 2^j divides a and let k be the largest integer such that 2^k divides b (note that $j \geq 0$ because $2^j = 2^0 = 1$ and 1 divides every integer). Since, 2^j divides a, there is $c \in \mathbb{Z}$ such that $a = 2^j c$. Since, 2^k divides b, there is $d \in \mathbb{Z}$ such that $b = 2^k d$.

Observe that c is odd. Indeed, if c were even, then there would be an integer s such that $c = 2s$. But then $a = 2^j c = 2^j (2s) = (2^j \cdot 2)s = (2^j \cdot 2^1)s = 2^{j+1} s$. So, 2^{j+1} divides a, contradicting the maximality of j.

Similarly, d is odd.

So, we have $x = \frac{a}{b} = \frac{2^j c}{2^k d}$.

If $j \geq k$, then, $j - k \geq 0$ and $x = \frac{2^j c}{2^k d} = \frac{2^{j-k} c}{d}$. Let $m = 2^{j-k} c$ and $n = d$. Then $x = \frac{m}{n}$, $m \in \mathbb{Z}$ (because $\mathbb{Z}$ is closed under multiplication), $n \in \mathbb{Z}^*$ (if $n = 0$, then $b = 2^k d = 2^k n = 2^k \cdot 0 = 0$, contradicting that $b \in \mathbb{Z}^*$), and $n = d$ is odd.

If $j < k$, then $k - j > 0$ and $x = \frac{2^j c}{2^k d} = \frac{c}{2^{k-j} d}$. Let $m = c$ and $n = 2^{k-j} d$. Then $x = \frac{m}{n}$, $m = c \in \mathbb{Z}$, $n \in \mathbb{Z}^*$ (because $\mathbb{Z}$ is closed under multiplication, and if n were 0, then d would be 0, and then b would be 0), and $m = c$ is odd. □

LEVEL 5

10. Show that every nonempty set of real numbers that is bounded below has a greatest lower bound in $\mathbb{R}$.

Proof: Let S be a nonempty set of real numbers that is bounded below. Let K be a lower bound of S, so that for all $x \in S$, $x \geq K$. Define the set T by $T = \{-x \mid x \in S\}$.

Let $y \in T$. Then there is $x \in S$ with $y = -x$. Since $x \in S$, $x \geq K$. It follows from Problem 5, part (vii) that $y = -x \leq -K$. Since $y \in T$ was arbitrary, we have shown that for all $y \in T$, $y \leq -K$. It follows that $-K$ is an upper bound of the set T.

By the Completeness Property of $\mathbb{R}$, T has a least upper bound M. We will show that $-M$ is a greatest lower bound of S.

Let $x \in S$. Then $-x \in T$. Since M is an upper bound of T, $-x \leq M$. So, by Problem 5, part (vii), $x \geq -M$. Since $x \in S$ was arbitrary, we have shown that for all $x \in S$, $x \geq -M$. Therefore, $-M$ is a lower bound of S.

Let $B > -M$. By Problem 5, part (vi), $-B < M$. Since M is the least upper bound of T, there is $y \in T$ with $y > -B$. By Problem 5, part (vi) again, we have $-y < B$. Since $y \in T$, $-y \in S$. Thus, B is not a lower bound of S.

Therefore, $-M$ is a greatest lower bound of S.

Since S was arbitrary, we have shown that every nonempty set of real numbers that is bounded below has a greatest lower bound in $\mathbb{R}$. □

> 11. Show that between any two real numbers there is a real number that is **not** rational.

Proof: Let $x, y \in \mathbb{R}$ with $x < y$. Let c be a positive number that is not rational. Then $\frac{x}{c} < \frac{y}{c}$. By the Density Theorem, there is a $q \in \mathbb{Q}$ such that $\frac{x}{c} < q < \frac{y}{c}$. We can assume that $q \neq 0$ (if it were, we could simply apply the Density Theorem again to get $p \in \mathbb{Q}$ with $\frac{x}{c} < p < q$, and p would not be 0). It follows that $x < cq < y$. Since $c = (cq)q^{-1}$, it follows that $cq \notin \mathbb{Q}$ (if $cq \in \mathbb{Q}$, then $c \in \mathbb{Q}$ because $\mathbb{Q}$ is closed under multiplication). So, cq is a real number between x and y that is **not** rational. □

> 12. Let $T = \{x \in F \mid -2 < x \leq 2\}$. Prove $\sup T = 2$ and $\inf T = -2$.

Proof: If $x \in T$, then by the definition of T, $x \leq 2$. So, 2 is an upper bound of T.

Now, let $B < 2$, and let $z = \max\{0, \frac{1}{2}(B+2)\}$. Since $B < 2$, we have

$$\tfrac{1}{2}(B+2) < \tfrac{1}{2}(2+2) = \tfrac{1}{2} \cdot 4 = 2.$$

So, if we have $\frac{1}{2}(B+2) > 0$, then $\frac{1}{2}(B+2) \in T$. Since $0 \in T$, we see that $z \in T$. Also,

$$z \geq \tfrac{1}{2}(B+2) > \tfrac{1}{2}(B+B) = \tfrac{1}{2}(2B) = \left(\tfrac{1}{2} \cdot 2\right)B = 1B = B.$$

So, we see that $z \in T$ and $z > B$. Therefore, B is not an upper bound of T. So, $2 = \sup T$.

If $x \in T$, then by the definition of T, $x > -2$. So, -2 is a lower bound of T.

Now, let $C > -2$, and let $w = \min\{0, \frac{1}{2}(-2+C)\}$. Since $C > -2$, we have

$$\tfrac{1}{2}(-2+C) > \tfrac{1}{2}(-2-2) = \tfrac{1}{2}(-4) = -2.$$

So, if we have $\frac{1}{2}(-2+C) < 0$, then $\frac{1}{2}(-2+C) \in T$. Since $0 \in T$, we see that $w \in T$. Also,

$$w \leq \tfrac{1}{2}(-2+C) < \tfrac{1}{2}(C+C) = \tfrac{1}{2}(2C) = \left(\tfrac{1}{2} \cdot 2\right)C = 1C = C.$$

So, we see that $w \in T$ and $w < C$. Therefore, C is not a lower bound of T. So, $-2 = \inf T$. □

Challenge Problem

13. Let $V = \{x \in F \mid x^2 < 2\}$ and let $a = \sup V$. Prove that $a^2 = 2$.

Hints:

- If $a^2 < 2$, find $n \in \mathbb{N}$ such that $a + \frac{1}{n} \in V$. This will contradict that a is an upper bound of V, proving $a^2 \geq 2$.
- If $a^2 > 2$, find $n \in \mathbb{N}$ such that $a - \frac{1}{n}$ is an upper bound of V. This will contradict that a is the least upper bound of V, proving $a^2 \leq 2$.
- You can use the Archimedean property of $\mathbb{R}$ with each of the last two hints.

Problem Set 6

LEVEL 1

1. Draw Venn diagrams for $(A \setminus B) \setminus C$ and $A \setminus (B \setminus C)$. Are these two sets equal for all sets A, B, and C? If so, prove it. If not, provide a counterexample.

Solution:

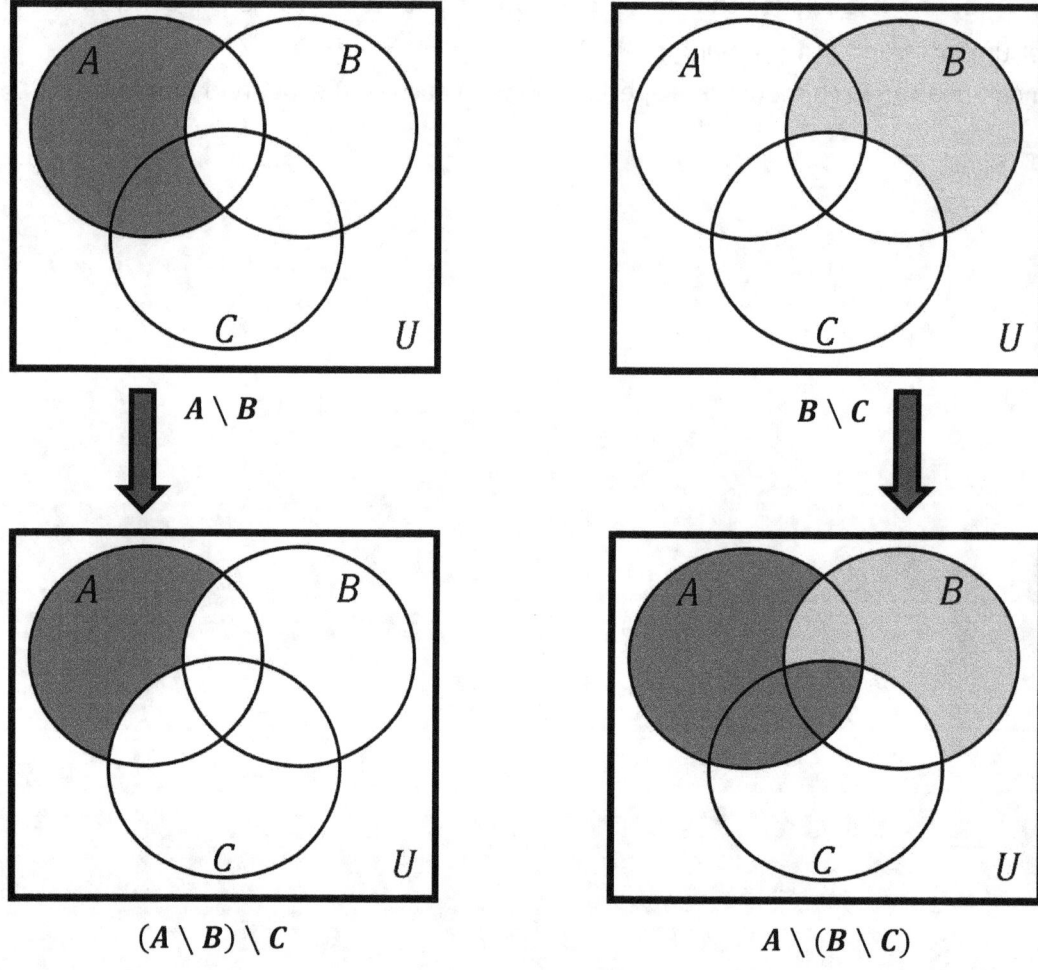

$A \setminus B$ $B \setminus C$

$(A \setminus B) \setminus C$ $A \setminus (B \setminus C)$

From the Venn diagrams, it looks like $(A \setminus B) \setminus C \subseteq A \setminus (B \setminus C)$, but $(A \setminus B) \setminus C \neq A \setminus (B \setminus C)$.

Let's come up with a counterexample. Let $A = \{1, 2\}$, $B = \{1, 3\}$, and $C = \{1, 4\}$. Then we have $(A \setminus B) \setminus C = \{2\} \setminus \{1, 4\} = \{2\}$ and $A \setminus (B \setminus C) = \{1, 2\} \setminus \{3\} = \{1, 2\}$.

We see that $(A \setminus B) \setminus C \neq A \setminus (B \setminus C)$.

Note: Although it was not asked in the question, let's prove that $(A \setminus B) \setminus C \subseteq A \setminus (B \setminus C)$. Let $x \in (A \setminus B) \setminus C$. Then $x \in A \setminus B$ and $x \notin C$. Since $x \in A \setminus B$, $x \in A$ and $x \notin B$. In particular, $x \in A$. Since $x \notin B$, $x \notin B \setminus C$ (because if $x \in B \setminus C$, then $x \in B$). So, we have $x \in A$ and $x \notin B \setminus C$. Therefore, $x \in A \setminus (B \setminus C)$. Since $x \in (A \setminus B) \setminus C$ was arbitrary, $(A \setminus B) \setminus C \subseteq A \setminus (B \setminus C)$. □

2. Let $A = \{\emptyset, \{\emptyset, \{\emptyset\}\}\}$, $B = \{\emptyset, \{\emptyset\}\}$, $C = (-\infty, 2]$, $D = (-1, 3]$. Compute each of the following:
(i) $A \cup B$; (ii) $A \cap B$; (iii) $A \setminus B$; (iv) $B \setminus A$; (v) $A \triangle B$; (vi) $C \cup D$; (vii) $C \cap D$; (viii) $C \setminus D$; (ix) $D \setminus C$; (x) $C \triangle D$.

Solutions:

(i) $A \cup B = \{\emptyset, \{\emptyset\}, \{\emptyset, \{\emptyset\}\}\}$

(ii) $A \cap B = \{\emptyset\}$

(iii) $A \setminus B = \{\{\emptyset, \{\emptyset\}\}\}$

(iv) $B \setminus A = \{\{\emptyset\}\}$

(v) $A \triangle B = \{\{\emptyset\}, \{\emptyset, \{\emptyset\}\}\}$

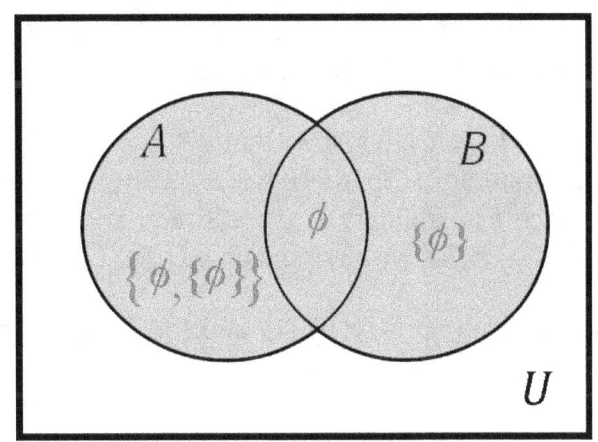

(vi) $C \cup D = (-\infty, 3]$

(vii) $C \cap D = (-1, 2]$

(viii) $C \setminus D = (-\infty, -1]$

(ix) $D \setminus C = (2, 3]$

(x) $C \triangle D = (-\infty, -1] \cup (2, 3]$

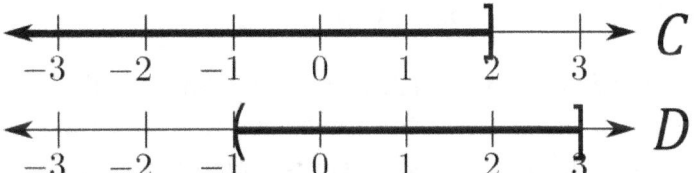

LEVEL 2

3. Prove the following: (i) The operation of forming unions is commutative. (ii) The operation of forming intersections is commutative. (iii) The operation of forming intersections is associative.

Proofs:

(i) Let A and B be sets. Then $x \in A \cup B$ if and only if $x \in A$ or $x \in B$ if and only if $x \in B$ or $x \in A$ if and only if $x \in B \cup A$. Since x was arbitrary, we have shown $\forall x (x \in A \cup B \leftrightarrow x \in B \cup A)$. Therefore, $A \cup B = B \cup A$. So, the operation of forming unions is commutative. □

(ii) Let A and B be sets. Then $x \in A \cap B$ if and only if $x \in A$ and $x \in B$ if and only if $x \in B$ and $x \in A$ if and only if $x \in B \cap A$. Since x was arbitrary, we have $\forall x (x \in A \cap B \leftrightarrow x \in B \cap A)$. Therefore, $A \cap B = B \cap A$. So, the operation of forming intersections is commutative. □

(iii) Let A, B, and C be sets. Then $x \in (A \cap B) \cap C$ if and only if $x \in A \cap B$ and $x \in C$ if and only if $x \in A$, $x \in B$ and $x \in C$ if and only if $x \in A$ and $x \in B \cap C$ if and only if $x \in A \cap (B \cap C)$. Since x was arbitrary, we have shown $\forall x (x \in (A \cap B) \cap C \leftrightarrow x \in A \cap (B \cap C))$.

Therefore, we have shown that $(A \cap B) \cap C = A \cap (B \cap C)$. So, the operation of forming intersections is associative. □

4. Prove that if an interval I is unbounded, then I has one of the following five forms: (a, ∞), $(-\infty, b)$, $[a, \infty)$, $(-\infty, b]$, $(-\infty, \infty)$

Proof: Let I be an unbounded interval. There are three cases to consider.

Case 1: I is bounded above, but not below. In this case, since I is bounded above, by the Completeness of $\mathbb{R}$, I has a least upper bound b. If $x \in I$, then by the definition of upper bound, we have $x \in (-\infty, b]$. Since x was an arbitrary element of I, $\forall x(x \in I \to x \in (-\infty, b])$. So, $I \subseteq (-\infty, b]$.

Now, let $z \in (-\infty, b)$. It follows that $z < b$. Since b is the **least** upper bound of I, it follows that z is **not** an upper bound of I. So, there is $y \in I$ with $z < y$. Since I is not bounded below, there is $x \in I$ with $x < z$. Since I is an interval, $x, y \in I$, and $x < z < y$, it follows that $z \in I$. Since z was an arbitrary element of $(-\infty, b)$, we have shown $\forall x(x \in (-\infty, b) \to x \in I)$. So, $(-\infty, b) \subseteq I$.

We have shown that $(-\infty, b) \subseteq I$ and $I \subseteq (-\infty, b]$. There are now 2 subcases to consider:

Subcase 1: If the least upper bound of I (namely, b) is an element of I, then we have $(-\infty, b] \subseteq I$ and $I \subseteq (-\infty, b]$. So, $I = (-\infty, b]$.

Subcase 2: If the least upper bound of I (namely, b) is not an element of I, then we have $(-\infty, b) \subseteq I$ and $I \subseteq (-\infty, b)$. So, $I = (-\infty, b)$.

Case 2: I is bounded below, but not above. In this case, since I is bounded below, by Problem 10 in Lesson 5, I has a greatest lower bound a. If $x \in I$, then by the definition of lower bound, we have $x \in [a, \infty)$. Since x was an arbitrary element of I, $\forall x(x \in I \to x \in [a, \infty))$. So, $I \subseteq [a, \infty)$.

Now, let $z \in (a, \infty)$. It follows that $z > a$. Since a is the **greatest** lower bound of I, it follows that z is **not** a lower bound of I. So, there is $x \in I$ with $x < z$. Since I is not bounded above, there is $y \in I$ with $z < y$. Since I is an interval, $x, y \in I$, and $x < z < y$, it follows that $z \in I$. Since z was an arbitrary element of (a, ∞), we have shown $\forall x(x \in (a, \infty) \to x \in I)$. So, $(a, \infty) \subseteq I$.

We have shown that $(a, \infty) \subseteq I$ and $I \subseteq [a, \infty)$. As in the last case, there are now 2 subcases to consider:

Subcase 1: If the greatest lower bound of I (namely, a) is an element of I, then we have $[a, \infty) \subseteq I$ and $I \subseteq [a, \infty)$. So, $I = [a, \infty)$.

Subcase 2: If the greatest lower bound of I (namely, a) is not an element of I, then we have $(a, \infty) \subseteq I$ and $I \subseteq (a, \infty)$. So, $I = (a, \infty)$.

Case 3: I is not bounded above or below. If $x \in I$, then $x \in \mathbb{R} = (-\infty, \infty)$. So, $I \subseteq (-\infty, \infty)$.

Now, let $z \in (-\infty, \infty)$. Since I is not bounded above, there is $y \in I$ with $z < y$. Since I is not bounded below, there is $x \in I$ with $x < z$. Since I is an interval, $x, y \in I$, and $x < z < y$, it follows that $z \in I$. Since z was an arbitrary element of $(-\infty, \infty)$, we have shown $\forall x(x \in (-\infty, \infty) \to x \in I)$. So, $(-\infty, \infty) \subseteq I$.

Since $I \subseteq (-\infty, \infty)$ and $(-\infty, \infty) \subseteq I$, we have $I = (-\infty, \infty)$ □

LEVEL 3

5. Prove or provide a counterexample: (i) Every pairwise disjoint set of sets is disjoint. (ii) Every disjoint set of sets is pairwise disjoint.

Solutions:

(i) This is **false**. Let $A = \{1\}$ and let $X = \{A\}$. X is pairwise disjoint, but $\cap X = A = \{1\} \neq \emptyset$.

However, the following slightly modified statement is **true**: "Every pairwise disjoint set of sets consisting of at least two sets is disjoint."

Let X be a pairwise disjoint set of sets with at least two sets, say $A, B \in X$. Suppose towards contradiction that $x \in \cap X$. Then $x \in A$ and $x \in B$. So, $x \in A \cap B$. But $A \cap B = \emptyset$ because X is pairwise disjoint. This contradiction shows that the statement $x \in \cap X$ is false. Therefore, X is disjoint. □

(ii) This is **false**. Let $A = \{0, 1\}$, $B = \{1, 2\}$, $C = \{0, 2\}$, and $X = \{A, B, C\}$. Then X is disjoint because $\cap X = A \cap B \cap C = \{0, 1\} \cap \{1, 2\} \cap \{0, 2\} = \{1\} \cap \{0, 2\} = \emptyset$. However, X is **not** pairwise disjoint because $A \cap B = \{0, 1\} \cap \{1, 2\} = \{1\} \neq \emptyset$.

6. Prove the following: (i) For all $b \in \mathbb{R}$, the infinite interval $(-\infty, b)$ is an open set in $\mathbb{R}$. (ii) The intersection of two open intervals in $\mathbb{R}$ is either empty or an open interval in $\mathbb{R}$. (iii) The intersection of finitely many open sets in $\mathbb{R}$ is an open set in $\mathbb{R}$.

Proofs:

(i) Let $x \in (-\infty, b)$ and let $a = x - 1$. Since $x \in (-\infty, b)$, $x < b$. Since $x - (x - 1) = 1 > 0$, we have $x > x - 1 = a$. So, we have $a < x < b$. That is, $x \in (a, b)$. Also, $(a, b) \subseteq (-\infty, b)$. Since $x \in (-\infty, b)$ was arbitrary, $(-\infty, b)$ is an open set. □

(ii) Let (a, b) and (c, d) be open intervals in $\mathbb{R}$ (a and c can be $-\infty$, and b and d can be ∞, where $-\infty$ is less than any real number and ∞, and ∞ is greater than any real number and $-\infty$). Without loss of generality, we may assume that $a \leq c$. If $b \leq c$, then we have $(a, b) \cap (c, d) = \emptyset$ because if $a < x < b$ and $c < x < d$, then $x < b \leq c < x$, and so, $x < x$, which is impossible.

So, we may assume that $c < b$. Let $e = \min\{b, d\}$. We claim that $(a, b) \cap (c, d) = (c, e)$.

Let $x \in (a, b) \cap (c, d)$. Then $x \in (a, b)$ and $x \in (c, d)$. So, $a < x < b$ and $c < x < d$. In particular, $x > c$, $x < b$, and $x < d$. Since $x < b$ and $x < d$, $x < e$. So, $x \in (c, e)$. Since $x \in (a, b) \cap (c, d)$ was arbitrary, we have shown that $(a, b) \cap (c, d) \subseteq (c, e)$.

Now, let $x \in (c, e)$. Then $c < x < e$. We are assuming that $a \leq c$. We also have $e \leq b$. So, $a \leq c < x < e \leq b$. Therefore, $x \in (a, b)$. We also have $e \leq d$. So, $c < x < e \leq d$, and therefore, $x \in (c, d)$. Since $x \in (a, b)$ and $x \in (c, d)$, we have $x \in (a, b) \cap (c, d)$. Since $x \in (c, e)$ was arbitrary, we have shown that $(c, e) \subseteq (a, b) \cap (c, d)$.

Finally, since we have shown $(a, b) \cap (c, d) \subseteq (c, e)$ and $(c, e) \subseteq (a, b) \cap (c, d)$, we have $(a, b) \cap (c, d) = (c, e)$.

Therefore, the intersection of two open intervals in $\mathbb{R}$ is either empty or an open interval in $\mathbb{R}$. □

(iii) The intersection of a single set with itself is just that set itself, and so, the result holds trivially for one open set.

So, we will prove the following statement: "The intersection of a set of finitely many open sets in $\mathbb{R}$ consisting of at least 2 sets is an open set in $\mathbb{R}$." We will prove this by induction on the number of open sets we are taking the intersection of. Theorem 6.9 is the base case $n = 2$.

For the inductive step, assume that the intersection of k nonempty open sets in $\mathbb{R}$ is open, and let X be a set of $k + 1$ open sets. Let $A \in X$ and let B be the intersection of all the sets in X except A. By the induction hypothesis, B is open. Therefore, $\cap X = A \cap B$ is open by Theorem 6.9.

By the Principle of Mathematical Induction, we have shown that the intersection of a set of finitely many open sets in $\mathbb{R}$ consisting of at least 2 sets is an open set in $\mathbb{R}$. □

7. Let $A, B,$ and C be sets. Prove each of the following: (i) $A \cap (B \cup C) = (A \cap B) \cup (A \cap C)$; (ii) $A \cup (B \cap C) = (A \cup B) \cap (A \cup C)$; (iii) $C \setminus (A \cup B) = (C \setminus A) \cap (C \setminus B)$; (iv) $C \setminus (A \cap B) = (C \setminus A) \cup (C \setminus B)$.

Proofs:

(i) $x \in A \cap (B \cup C) \Leftrightarrow x \in A$ and $x \in B \cup C \Leftrightarrow x \in A$ and either $x \in B$ or $x \in C \Leftrightarrow x \in A$ and $x \in B$ or $x \in A$ and $x \in C \Leftrightarrow x \in A \cap B$ or $x \in A \cap C \Leftrightarrow x \in (A \cap B) \cup (A \cap C)$. □

(ii) $x \in A \cup (B \cap C) \Leftrightarrow x \in A$ or $x \in B \cap C \Leftrightarrow$ either $x \in A$ or we have both $x \in B$ and $x \in C \Leftrightarrow$ we have both $x \in A$ or $x \in B$ and $x \in A$ or $x \in C \Leftrightarrow x \in A \cup B$ and $x \in A \cup C \Leftrightarrow x \in (A \cup B) \cap (A \cup C)$. □

(iii) $x \in C \setminus (A \cup B) \Leftrightarrow x \in C$ and $x \notin A \cup B \Leftrightarrow x \in C$ and $x \notin A$ and $x \notin B \Leftrightarrow x \in C$ and $x \notin A$ and $x \in C$ and $x \notin B \Leftrightarrow x \in C \setminus A$ and $x \in C \setminus B \Leftrightarrow x \in (C \setminus A) \cap (C \setminus B)$. □

(iv) $x \in C \setminus (A \cap B) \Leftrightarrow x \in C$ and $x \notin A \cap B \Leftrightarrow x \in C$ and either $x \notin A$ or $x \notin B \Leftrightarrow x \in C$ and $x \notin A$ or $x \in C$ and $x \notin B \Leftrightarrow x \in C \setminus A$ or $x \in C \setminus B \Leftrightarrow x \in (C \setminus A) \cup (C \setminus B)$. □

Notes: Let's let $p, q,$ and r be the statements $x \in A, x \in B,$ and $x \in C$, respectively.

(1) In (i) above, the statement "$x \in A$ and either $x \in B$ or $x \in C$" can be written $p \wedge (q \vee r)$. By Problem 9 in Lesson 1, this is equivalent to $(p \wedge q) \vee (p \wedge r)$. In words, this is the statement "$x \in A$ and $x \in B$ or $x \in A$ and $x \in C$." Here it needs to be understood that the word "and" takes precedence over the word "or."

Similarly, we can use the logical equivalence $p \vee (q \wedge r) \equiv (p \vee q) \wedge (p \vee r)$ to help understand the proof of (ii).

(2) Recall that the equivalences $p \wedge (q \vee r) \equiv (p \wedge q) \vee (p \wedge r)$ and $p \vee (q \wedge r) \equiv (p \vee q) \wedge (p \vee r)$ are known as the **distributive laws** (see Note (4) following the solution to Problem 9 in Problem Set 1).

The rules $A \cap (B \cup C) = (A \cap B) \cup (A \cap C)$ and $A \cup (B \cap C) = (A \cup B) \cap (A \cup C)$ are also known as the **distributive laws**.

(3) To clarify (iii) and (iv), recall the **De Morgan's laws** $\neg(p \lor q) \equiv \neg p \land \neg q$ and $\neg(p \land q) \equiv \neg p \lor \neg q$ from the notes following the solutions to Problem 3 from Problem Set 1. For (iii), we can use the logical equivalence $\neg(p \lor q) \equiv \neg p \land \neg q$ with p the statement $x \in A$ and q the statement $x \in B$ to get

$$x \notin A \cup B \equiv \neg x \in A \cup B \equiv \neg(x \in A \lor x \in B) \equiv \neg(p \lor q) \equiv \neg p \land \neg q \text{ (by De Morgan's law)}$$
$$\equiv \neg x \in A \land \neg x \in B \equiv x \notin A \land x \notin B.$$

So, the statement "$x \in C$ and $x \notin A \cap B$" is equivalent to $x \in C \land x \notin A \land x \notin B$.

Similarly, we can use the logical equivalence $\neg(p \land q) \equiv \neg p \lor \neg q$ to see that the statement "$x \in C$ and $x \notin A \cap B$" is equivalent to "$x \in C$ and $x \notin A$ or $x \notin B$."

(4) The rules $C \setminus (A \cup B) = (C \setminus A) \cap (C \setminus B)$ and $C \setminus (A \cap B) = (C \setminus A) \cup (C \setminus B)$ are also known as **De Morgan's laws**.

LEVEL 4

8. Give an example of an infinite collection of open sets whose intersection is not open. Also, give an example of an infinite collection of closed sets whose union is not closed. Provide a proof for each example.

Solution: Let $X = \left\{ \left(0, 1 + \frac{1}{n}\right) \mid n \in \mathbb{Z}^+ \right\}$. Each set in X is an open interval, and therefore, open. We will show that $\cap X = (0,1]$. Note that $x \in \cap X$ if and only if for all $n \in \mathbb{Z}^+$, $x \in \left(0, 1 + \frac{1}{n}\right)$ if and only if for all $n \in \mathbb{Z}^+$, $0 < x < 1 + \frac{1}{n}$. We need to show that $x \leq 1$ is equivalent to $\forall n \in \mathbb{Z}^+ \left(x < 1 + \frac{1}{n}\right)$.

Suppose that $x \leq 1$. Let $n \in \mathbb{Z}^+$. By Theorem 5.4, $\frac{1}{n} > 0$. So, $1 + \frac{1}{n} - 1 > 0$ (SACT). Thus, $1 + \frac{1}{n} > 1$. So, we have $x \leq 1 < 1 + \frac{1}{n}$, and therefore, $x < 1 + \frac{1}{n}$. Since $n \in \mathbb{Z}^+$ was arbitrary, we have shown that $x \leq 1$ implies $\forall n \in \mathbb{Z}^+ \left(x < 1 + \frac{1}{n}\right)$.

Now, suppose $x > 1$ (proof by contrapositive). Then $x - 1 > 0$. Since there is no smallest positive real number (by Problem 7 from Problem Set 5), there is a real number $\epsilon > 0$ with $x - 1 > \epsilon$. By the Archimedean Property of the reals, there is a natural number n with $n > \frac{1}{\epsilon}$. So, $\frac{1}{n} < \epsilon$, or equivalently, $\epsilon > \frac{1}{n}$. Thus, $x - 1 > \frac{1}{n}$, and so, $x > 1 + \frac{1}{n}$. We have shown that there is $n \in \mathbb{Z}^+$ such that $x > 1 + \frac{1}{n}$. So, $\forall n \in \mathbb{Z}^+ \left(x < 1 + \frac{1}{n}\right)$ is false.

This equivalence proves that $\cap X = (0, 1]$, an interval that is **not** an open set.

Let $Y = \left\{ \left[0, 1 - \frac{1}{n}\right] \mid n \in \mathbb{Z}^+ \right\}$. Each set in Y is a closed interval, and therefore, closed. We will show that $\cup Y = [0,1)$. Note that $x \in \cup Y$ if and only if there is $n \in \mathbb{Z}^+$ such that $x \in \left[0, 1 - \frac{1}{n}\right]$ if and only if there is $n \in \mathbb{Z}^+$ such that $0 \leq x \leq 1 - \frac{1}{n}$. We need to show that $x < 1$ is equivalent to $\exists n \in \mathbb{Z}^+ \left(x \leq 1 - \frac{1}{n}\right)$ (where $\exists$ is read "there exists" or "there is").

Suppose $x < 1$. Then $1 - x > 0$. Since there is no smallest positive real number (by Problem 7 from Problem Set 5), there is a real number $\epsilon > 0$ with $1 - x > \epsilon$. By the Archimedean Property of the reals, there is a natural number n with $n > \frac{1}{\epsilon}$. So, $\frac{1}{n} < \epsilon$, or equivalently, $\epsilon > \frac{1}{n}$. Thus, $1 - x > \frac{1}{n}$, and so, $x < 1 - \frac{1}{n}$. It follows that $x \leq 1 - \frac{1}{n}$. We have shown that there is $n \in \mathbb{Z}^+$ such that $x \leq 1 - \frac{1}{n}$. So, $\exists n \in \mathbb{Z}^+ \left(x \leq 1 - \frac{1}{n}\right)$.

Now, suppose $x \geq 1$ (proof by contrapositive). Let $n \in \mathbb{Z}^+$. By Theorem 5.4, $\frac{1}{n} > 0$. So, $1 - 1 + \frac{1}{n} > 0$ (SACT). So, $1 > 1 - \frac{1}{n}$. It follows that $x > 1 - \frac{1}{n}$. Since $n \in \mathbb{Z}^+$ was arbitrary, $\forall n \in \mathbb{Z}^+ \left(x > 1 - \frac{1}{n}\right)$. It follows that $\exists n \in \mathbb{Z}^+ \left(x \leq 1 - \frac{1}{n}\right)$ is false.

This equivalence proves that $\bigcup Y = [0, 1)$, an interval that is **not** a closed set.

9. Let X be a nonempty set of sets. Prove the following: (i) For all $A \in X$, $A \subseteq \bigcup X$. (ii) For all $A \in X$, $\bigcap X \subseteq A$.

Proofs:

(i) Let X be a nonempty set of sets, let $A \in X$, and let $x \in A$. Then there is $B \in X$ such that $x \in B$ (namely A). So, $x \in \bigcup X$. Since x was an arbitrary element of A, we have shown that $A \subseteq \bigcup X$. Since A was an arbitrary element of X, we have shown that for all $A \in X$, we have $A \subseteq \bigcup X$. □

(ii) Let X be a nonempty set of sets, let $A \in X$, and let $x \in \bigcap X$. Then for every $B \in X$, we have $x \in B$. In particular, $x \in A$ (because $A \in X$). Since x was an arbitrary element of $\bigcap X$, we have shown that $\bigcap X \subseteq A$. Since A was an arbitrary element of X, we have shown that for all $A \in X$, we have $\bigcap X \subseteq A$. □

LEVEL 5

10. Prove that if X is a nonempty set of closed subsets of $\mathbb{R}$, then $\bigcap X$ is closed.

Proof: Let X be a nonempty set of closed subsets of $\mathbb{R}$. Then for each $A \in X$, $\mathbb{R} \setminus A$ is an open set in $\mathbb{R}$. By Theorem 6.7, $\bigcup \{\mathbb{R} \setminus A \mid A \in X\}$ is open in $\mathbb{R}$. Therefore, $\mathbb{R} \setminus \bigcup \{\mathbb{R} \setminus A \mid A \in X\}$ is closed in $\mathbb{R}$. So, it suffices to show that $\bigcap X = \mathbb{R} \setminus \bigcup \{\mathbb{R} \setminus A \mid A \in X\}$. Well, $x \in \bigcap X$ if and only if for all $A \in X$, $x \in A$ if and only if for all $A \in X$, $x \notin \mathbb{R} \setminus A$ if and only if $x \notin \bigcup \{\mathbb{R} \setminus A \mid A \in X\}$ if and only if $x \in \mathbb{R} \setminus \bigcup \{\mathbb{R} \setminus A \mid A \in X\}$. So, $\bigcap X = \mathbb{R} \setminus \bigcup \{\mathbb{R} \setminus A \mid A \in X\}$, completing the proof. □

11. Let A be a set and let X be a nonempty collection of sets. Prove each of the following:
    (i) $A \cap \bigcup X = \bigcup \{A \cap B \mid B \in X\}$; (ii) $A \cup \bigcap X = \bigcap \{A \cup B \mid B \in X\}$;
    (iii) $A \setminus \bigcup X = \bigcap \{A \setminus B \mid B \in X\}$; (iv) $A \setminus \bigcap X = \bigcup \{A \setminus B \mid B \in X\}$.

Proofs:

(i) $x \in A \cap \bigcup X \Leftrightarrow x \in A$ and $x \in \bigcup X \Leftrightarrow x \in A$ and there is a $B \in X$ with $x \in B \Leftrightarrow x \in A \cap B$ for some $B \in X \Leftrightarrow x \in \bigcup \{A \cap B \mid B \in X\}$. □

(ii) $x \in A \cup \bigcap X \Leftrightarrow x \in A$ or $x \in \bigcap X \Leftrightarrow x \in A$ or $x \in B$ for every $B \in X \Leftrightarrow x \in A \cup B$ for every $B \in X \Leftrightarrow x \in \bigcap\{A \cup B \mid B \in X\}$. □

(iii) $x \in A \setminus \bigcup X \Leftrightarrow x \in A$ and $x \notin \bigcup X \Leftrightarrow x \in A$ and $x \notin B$ for every $B \in X \Leftrightarrow x \in A \setminus B$ for every $B \in X \Leftrightarrow x \in \bigcap\{A \setminus B \mid B \in X\}$. □

(iv) $x \in A \setminus \bigcap X \Leftrightarrow x \in A$ and $x \notin \bigcap X \Leftrightarrow x \in A$ and $x \notin B$ for some $B \in X \Leftrightarrow x \in A \setminus B$ for some $B \in X \Leftrightarrow x \in \bigcup\{A \setminus B \mid B \in X\}$. □

Note: The rules in (i) and (ii) are known as the **generalized distributive laws** and the rules in (iii) and (iv) are known as the **generalized De Morgan's laws.**

12. Prove that every closed set in $\mathbb{R}$ can be written as an intersection $\bigcap X$, where each element of X is a union of at most 2 closed intervals.

Proof: First note that $\mathbb{R} = \bigcap\{\mathbb{R}\}$.

Let A be a closed set in $\mathbb{R}$ with $A \neq \mathbb{R}$. Then $\mathbb{R} \setminus A$ is a nonempty open set in $\mathbb{R}$. By Theorem 6.8, $\mathbb{R} \setminus A$ can be expressed as $\bigcup X$, where X is a set of bounded open intervals. For each B in X, $\mathbb{R} \setminus B$ is a union of two closed intervals (if $B = (a, b)$, then $\mathbb{R} \setminus B = (-\infty, a] \cup [b, \infty)$). Now, by part (iii) of Problem 11, we have $A = \mathbb{R} \setminus (\mathbb{R} \setminus A) = \mathbb{R} \setminus \bigcup X = \bigcap\{\mathbb{R} \setminus B \mid B \in X\}$. □

Problem Set 7

LEVEL 1

1. Let $z = -4 - i$ and $w = 3 - 5i$. Compute each of the following: (i) $z + w$; (ii) zw; (iii) Im w; (iv) $2z - w$; (v) $\overline{w}$; (vi) $\frac{z}{w}$; (vii) $|z|$; (viii) the distance between z and w

Solutions:

(i) $z + w = (-4 - i) + (3 - 5i) = (-4 + 3) + (-1 - 5)i = \mathbf{-1 - 6i}$.

(ii) $zw = (-4 - i)(3 - 5i) = (-12 - 5) + (20 - 3)i = \mathbf{-17 + 17i}$.

(iii) Im w = Im $(3 - 5i) = \mathbf{-5}$.

(iv) $2z - w = 2(-4 - i) - (3 - 5i) = (-8 - 2i) + (-3 + 5i) = (-8 - 3) + (-2 + 5)i$
$= \mathbf{-11 + 3i}$.

(v) $\overline{w} = \overline{3 - 5i} = \mathbf{3 + 5i}$.

(vi) $\frac{z}{w} = \frac{-4-i}{3-5i} = \frac{(-4-i)(3+5i)}{(3-5i)(3+5i)} = \frac{(-12+5)+(-20-3)i}{3^2+5^2} = \frac{-7-23i}{9+25} = \mathbf{-\frac{7}{34} - \frac{23}{34}i}$.

(vii) $|z| = \sqrt{(-4)^2 + (-1)^2} = \sqrt{16 + 1} = \mathbf{\sqrt{17}}$.

(viii) $|z - w| = |(-4 - i) - (3 - 5i)| = |(-4 - 3) + (-1 + 5)i| = |-7 + 4i| = \sqrt{(-7)^2 + 4^2}$
$= \sqrt{49 + 16} = \mathbf{\sqrt{65}}$.

LEVEL 2

2. Prove that $(\mathbb{C}, +, \cdot)$ is field.

Solution: We first prove that $(\mathbb{C}, +)$ is a commutative group.

(Closure) Let $z, w \in \mathbb{C}$. Then there are $a, b, c, d \in \mathbb{R}$ such that $z = a + bi$ and $w = c + di$. By definition, $z + w = (a + bi) + (c + di) = (a + c) + (b + d)i$. Since $\mathbb{R}$ is closed under addition, $a + b \in \mathbb{R}$ and $c + d \in \mathbb{R}$. Therefore, $z + w \in \mathbb{C}$.

(Associativity) Let $z, w, v \in \mathbb{C}$. Then there are $a, b, c, d, e, f \in \mathbb{R}$ such that $z = a + bi, w = c + di$, and $v = e + fi$. Since addition is associative in $\mathbb{R}$, we have

$(z + w) + v = \big((a + bi) + (c + di)\big) + (e + fi) = \big((a + c) + (b + d)i\big) + (e + fi)$
$= \big((a + c) + e\big) + \big((b + d) + f\big)i = \big(a + (c + e)\big) + \big(b + (d + f)\big)i$
$= (a + bi) + \big((c + e) + (d + f)i\big) = (a + bi) + \big((c + di) + (e + fi)\big) = z + (w + v)$.

(Commutativity) Let $z, w \in \mathbb{C}$. Then there are $a, b, c, d \in \mathbb{R}$ such that $z = a + bi$ and $w = c + di$. Since addition is commutative in $\mathbb{R}$, we have

$z + w = (a + bi) + (c + di) = (a + c) + (b + d)i = (c + a) + (d + b)i$
$= (c + di) + (a + bi) = w + z$.

(Identity) Let $\overline{0} = 0 + 0i$. We show that $\overline{0}$ is an additive identity for $\mathbb{C}$. Since $0 \in \mathbb{R}$, $\overline{0} \in \mathbb{C}$. Let $z \in \mathbb{C}$. Then there are $a, b \in \mathbb{R}$ such that $z = a + bi$. Since 0 is an additive identity in $\mathbb{R}$, we have

$$\overline{0} + z = (0 + 0i) + (a + bi) = (0 + a) + (0 + b)i = a + bi.$$
$$z + \overline{0} = (a + bi) + (0 + 0i) = (a + 0) + (b + 0)i = a + bi.$$

(Inverse) Let $z \in \mathbb{C}$. Then there are $a, b \in \mathbb{R}$ such that $z = a + bi$. Let $w = -a + (-b)i$. Then

$$z + w = (a + bi) + (-a + (-b)i) = (a + (-a)) + (b + (-b))i = 0 + 0i = \overline{0}.$$
$$w + z = (-a + (-b)i) + (a + bi) = (-a + a) + (-b + b)i = 0 + 0i = \overline{0}.$$

We next prove that $(\mathbb{C}^*, \cdot)$ is a commutative group.

(Closure) Let $z, w \in \mathbb{C}^*$. Then there are $a, b, c, d \in \mathbb{R}$ such that $z = a + bi$ and $w = c + di$. By definition, $zw = (a + bi)(c + di) = (ac - bd) + (ad + bc)i$. Since $\mathbb{R}$ is closed under multiplication, we have $ac, bd, ad, bc \in \mathbb{R}$. Also, $-bd$ is the additive inverse of bd in $\mathbb{R}$. Since $\mathbb{R}$ is closed under addition, we have $ac - bd = ac + (-bd) \in \mathbb{R}$ and $ad + bc \in \mathbb{R}$. Therefore, $zw \in \mathbb{C}$.

We still need to show that $zw \neq 0$. If $zw = 0$, then $ac - bd = 0$ and $ad + bc = 0$. So, $ac = bd$ and $ad = -bc$. Multiplying each side of the last equation by c gives us $acd = -bc^2$. Replacing ac with bd on the left gives $bd^2 = -bc^2$, or equivalently, $bd^2 + bc^2 = 0$. So, $b(d^2 + c^2) = 0$. If $d^2 + c^2 = 0$, then $c = 0$ and $d = 0$, and so, $w = 0$. If $b = 0$, then $ac = 0$, and so, $a = 0$ or $c = 0$. If $a = 0$, then $z = 0$. If $c = 0$ and $a \neq 0$, then since $ad = -bc = 0$, we have $d = 0$. So, $w = 0$. So, we see that $zw = 0$ implies $z = 0$ or $w = 0$. By contrapositive, since $z, w \in \mathbb{C}^*$, we must have $zw \neq 0$, and so, $zw \in \mathbb{C}^*$.

(Associativity) Let $z, w, v \in \mathbb{C}^*$. Then there are $a, b, c, d, e, f \in \mathbb{R}$ such that $z = a + bi$, $w = c + di$, and $v = e + fi$. Since addition and multiplication are associative in $\mathbb{R}$, addition is commutative in $\mathbb{R}$, and multiplication is distributive over addition in $\mathbb{R}$, we have

$$(zw)v = ((a + bi)(c + di))(e + fi) = ((ac - bd) + (ad + bc)i)(e + fi)$$
$$= [(ac - bd)e - (ad + bc)f] + [(ac - bd)f + (ad + bc)e]i$$
$$= (ace - bde - adf - bcf) + (acf - bdf + ade + bce)i$$
$$= (ace - adf - bcf - bde) + (acf + ade + bce - bdf)i$$
$$= [a(ce - df) - b(cf + de)] + [a(cf + de) + b(ce - df)]i$$
$$= (a + bi)((ce - df) + (cf + de)i) = (a + bi)((c + di)(e + fi)) = z(wv).$$

(Commutativity) Let $z, w \in \mathbb{C}^*$. Then there are $a, b, c, d \in \mathbb{R}$ such that $z = a + bi$ and $w = c + di$. Since addition and multiplication are commutative in $\mathbb{R}$, we have

$$zw = (a + bi)(c + di) = (ac - bd) + (ad + bc)i$$
$$= (ca - db) + (cb + da)i = (c + di)(a + bi) = wz$$

(Identity) Let $\overline{1} = 1 + 0i$. We show that $\overline{1}$ is a multiplicative identity for $\mathbb{C}^*$. Since $0, 1 \in \mathbb{R}$, $\overline{1} \in \mathbb{C}^*$. Let $z \in \mathbb{C}^*$. Then there are $a, b \in \mathbb{R}$ such that $z = a + bi$. Since 0 is an additive identity in $\mathbb{R}$, 1 is a multiplicative identity in $\mathbb{R}$, and $0 \cdot x = x \cdot 0 = 0$ for all $x \in \mathbb{R}$, we have

$$\bar{1}z = (1 + 0i)(a + bi) = (1a - 0b) + (1b + 0a)i = 1a + 1bi = a + bi.$$
$$z \cdot \bar{1} = (a + bi)(1 + 0i) = (a \cdot 1 - b \cdot 0) + (a \cdot 0 + b \cdot 1)i = a \cdot 1 + b \cdot 1i = a + bi.$$

(Inverse) Let $z \in \mathbb{C}^*$. Then there are $a, b \in \mathbb{R}$ such that $z = a + bi$. Let $w = \frac{a}{a^2+b^2} + \frac{-b}{a^2+b^2}i$. Then we have

$$zw = (a + bi)\left(\frac{a}{a^2 + b^2} + \frac{-b}{a^2 + b^2}i\right)$$
$$= \left(a \cdot \frac{a}{a^2 + b^2} - b \cdot \frac{-b}{a^2 + b^2}\right) + \left(a \cdot \frac{-b}{a^2 + b^2} + b \cdot \frac{a}{a^2 + b^2}\right)i$$
$$= \frac{a^2 + b^2}{a^2 + b^2} + \frac{-ab + ba}{a^2 + b^2}i = 1 + 0i = \bar{1}.$$

$$wz = \left(\frac{a}{a^2 + b^2} + \frac{-b}{a^2 + b^2}i\right)(a + bi)$$
$$= \left(\frac{a}{a^2 + b^2} \cdot a - \frac{-b}{a^2 + b^2} \cdot b\right) + \left(\frac{a}{a^2 + b^2} \cdot b + \frac{-b}{a^2 + b^2} \cdot a\right)i$$
$$= \frac{a^2 + b^2}{a^2 + b^2} + \frac{ab - ba}{a^2 + b^2}i = 1 + 0i = \bar{1}.$$

(Left Distributivity) Let $z, w, v \in \mathbb{C}$. Then there are $a, b, c, d, e, f \in \mathbb{R}$ such that $z = a + bi$, $w = c + di$, and $v = e + fi$. Since multiplication is left distributive over addition in $\mathbb{R}$, and addition is associative and commutative in $\mathbb{R}$, we have

$$z(w + v) = (a + bi)[(c + di) + (e + fi)] = (a + bi)[(c + e) + (d + f)i]$$
$$= [a(c + e) - b(d + f)] + [a(d + f) + b(c + e)]i$$
$$= (ac + ae - bd - bf) + (ad + af + bc + be)i$$
$$= [(ac - bd) + (ad + bc)i] + [(ae - bf) + (af + be)i]$$
$$(a + bi)(c + di) + (a + bi)(e + fi) = zw + zv.$$

(Right Distributivity) Let $z, w, v \in \mathbb{C}$. There are $a, b, c, d, e, f \in \mathbb{R}$ such that $z = a + bi$, $w = c + di$, and $v = e + fi$. Since multiplication is right distributive over addition in $\mathbb{R}$, and addition is associative and commutative in $\mathbb{R}$, we have

$$(w + v)z = [(c + di) + (e + fi)](a + bi) = [(c + e) + (d + f)i](a + bi)$$
$$= [(c + e)a - (d + f)b] + [(c + e)b + (d + f)a]i$$
$$= (ca + ea - db - fb) + (cb + eb + da + fa)i$$
$$= [(ca - db) + (cb + da)i] + [(ea - fb) + (eb + fa)i]$$
$$(c + di)(a + bi) + (e + fi)(a + bi) = wz + vz.$$

Therefore, $(\mathbb{C}, +, \cdot)$ is a field. □

Note: When verifying the inverse property, we didn't mention the field properties that were used and we skipped some steps. The dedicated reader may want to fill in these details.

3. Let z and w be complex numbers. Prove the following: (i) $\operatorname{Re} z = \frac{z+\bar{z}}{2}$; (ii) $\operatorname{Im} z = \frac{z-\bar{z}}{2i}$ (iii) $\overline{z+w} = \bar{z}+\bar{w}$; (iv) $\overline{zw} = \bar{z}\cdot\bar{w}$; (v) $\overline{\left(\frac{z}{w}\right)} = \frac{\bar{z}}{\bar{w}}$; (vi) $z\bar{z} = |z|^2$; (vii) $|zw| = |z||w|$; (viii) If $w \neq 0$, then $\left|\frac{z}{w}\right| = \frac{|z|}{|w|}$; (ix) $\operatorname{Re} z \leq |z|$; (x) $\operatorname{Im} z \leq |z|$

Proofs:

(i) Let $z = a + bi$. Then $\bar{z} = a - bi$, and so, $\frac{z+\bar{z}}{2} = \frac{(a+bi)+(a-bi)}{2} = \frac{2a+bi-bi}{2} = \frac{2a}{2} = a = \operatorname{Re} z$. □

(ii) Let $z = a + bi$. Then $\frac{z-\bar{z}}{2i} = \frac{(a+bi)-(a-bi)}{2i} = \frac{a-a+bi+bi}{2i} = \frac{2bi}{2i} = b = \operatorname{Im} z$. □

(iii) Let $z = a + bi$ and $w = c + di$. Then we have
$$\overline{z+w} = \overline{(a+bi)+(c+di)} = \overline{(a+c)+(b+d)i} = (a+c)-(b+d)i$$
$$= (a-bi)+(c-di) = \overline{a+bi}+\overline{c+di} = \bar{z}+\bar{w}.$$ □

(iv) Let $z = a + bi$ and $w = c + di$. Then we have
$$\overline{zw} = \overline{(a+bi)(c+di)} = \overline{(ac-bd)+(ad+bc)i} = (ac-bd)-(ad+bc)i$$
$$= (a-bi)(c-di) = \overline{(a+bi)}\,\overline{(c+di)} = \bar{z}\cdot\bar{w}.$$ □

(v) Let $z = a + bi$ and $w = c + di$. Then we have
$$\overline{\left(\frac{z}{w}\right)} = \overline{\frac{(a+bi)}{(c+di)}} = \overline{\frac{(a+bi)}{(c+di)}\cdot\frac{(c-di)}{(c-di)}} = \overline{\frac{(a+bi)(c-di)}{(c+di)(c-di)}} = \overline{\frac{(ac+bd)+(-ad+bc)i}{c^2+d^2}}$$
$$= \overline{\frac{(ac+bd)}{c^2+d^2} + \frac{(-ad+bc)}{c^2+d^2}i} = \frac{(ac+bd)}{c^2+d^2} - \frac{(-ad+bc)}{c^2+d^2}i = \frac{(ac+bd)+(ad-bc)i}{c^2+d^2}$$
$$= \frac{(a-bi)(c+di)}{(c-di)(c+di)} = \frac{(a-bi)}{(c-di)}\cdot\frac{(c+di)}{(c+di)} = \frac{a-bi}{c-di} = \frac{\overline{a+bi}}{\overline{c+di}} = \frac{\bar{z}}{\bar{w}}$$ □

(vi) Let $z = a + bi$. Then $z\bar{z} = (a+bi)(a-bi) = a^2+b^2 = |z|^2$. □

(vii) Using associativity and commutativity of multiplication in $\mathbb{C}$, as well as (iv) (vi) above, we have $|zw|^2 = (zw)(\overline{zw}) = (zw)(\bar{z}\cdot\bar{w}) = z\bar{z}w\bar{w} = |z|^2|w|^2 = (|z||w|)^2$. Since $|z|, |w|$, and $|zw|$ are nonnegative, $|zw| = |z||w|$. □

(viii) Using (vii) above, we have $\left|\frac{z}{w}\right| = |zw^{-1}| = |z||w^{-1}| = \frac{|z|}{|w|}$. □

(ix) We have $|z|^2 = (\operatorname{Re} z)^2 + (\operatorname{Im} z)^2$. Therefore, $(\operatorname{Re} z)^2 \leq |z|^2$, and so, $|\operatorname{Re} z| \leq |z|$. It follows that $\operatorname{Re} z \leq |\operatorname{Re} z| \leq |z|$. □

(x) We have $|z|^2 = (\operatorname{Re} z)^2 + (\operatorname{Im} z)^2$. Therefore, $(\operatorname{Im} z)^2 \leq |z|^2$, and so, $|\operatorname{Im} z| \leq |z|$. It follows that $\operatorname{Im} z \leq |\operatorname{Im} z| \leq |z|$. □

LEVEL 3

4. Prove the Triangle Inequality (Theorem 7.3).

Proof: $|z+w|^2 = (z+w)\overline{(z+w)} = (z+w)(\bar{z}+\bar{w}) = z\bar{z} + z\bar{w} + w\bar{z} + w\bar{w}$
$= z\bar{z} + z\bar{w} + \overline{z\bar{w}} + w\bar{w} = z\bar{z} + 2\text{Re}(z\bar{w}) + w\bar{w} \leq z\bar{z} + 2|z\bar{w}| + w\bar{w} = z\bar{z} + 2|z||\bar{w}| + w\bar{w}$
$= |z|^2 + 2|z||w| + |w|^2 = (|z| + |w|)^2$

Since $|z+w|$ and $|z|+|w|$ are nonnegative, $|z+w| \leq |z|+|w|$. □

5. Let z and w be complex numbers. Prove $||z| - |w|| \leq |z \pm w| \leq |z| + |w|$.

Proof: $|z| = |(z+w) + (-w)| \leq |z+w| + |-w| = |z+w| + |w|$. So, $|z+w| \geq |z| - |w|$.

$|w| = |(z+w) + (-z)| \leq |z+w| + |-z| = |z+w| + |z|$. So, $|z+w| \geq |w| - |z| = -(|z| - |w|)$.

Since for all $w, z \in \mathbb{C}$, we have $||z| - |w|| = |z| - |w|$ or $||z| - |w|| = -(|z| - |w|)$, it follows that $||z| - |w|| \leq |z+w|$.

Combining this result with the Triangle Inequality, gives us $||z| - |w|| \leq |z+w| \leq |z| + |w|$.

Now, by the Triangle Inequality we have $|z - w| = |z + (-w)| \leq |z| + |-w| = |z| + |w|$.

Finally, by the third paragraph, we have $|z - w| = |z + (-w)| \geq ||z| - |-w|| = ||z| - |w||$. □

6. A point w is an **accumulation point** of a set S of complex numbers if each deleted neighborhood of w contains at least one point in S. Determine the accumulation points of each of the following sets: (i) $\{\frac{1}{n} | n \in \mathbb{Z}^+\}$; (ii) $\{\frac{i}{n} | n \in \mathbb{Z}^+\}$; (iii) $\{i^n | n \in \mathbb{Z}^+\}$; (iv) $\{\frac{i^n}{n} | n \in \mathbb{Z}^+\}$; (v) $\{z | |z| < 1\}$; (vi) $\{z | 0 < |z - 2| \leq 3\}$

Solutions:

(i) **0** is the only accumulation point of this set.

(ii) **0** is the only accumulation point of this set.

(iii) This set is equal to $\{1, -1, i, -i\}$. It has **no accumulation points**.

(iv) **0** is the only accumulation point of this set.

(v) The set of accumulation points of the set $\{z | |z| < 1\}$ is the set $\{z \,|\, |z| \leq 1\}$.

(vi) The set of accumulation points of the set $\{z | 0 < |z - 2| \leq 3\}$ is the set $\{z | |z - 2| \leq 3\}$.

LEVEL 4

7. Determine if each of the following subsets of $\mathbb{C}$ is open, closed, both, or neither. Give a proof in each case. (i) $\emptyset$; (ii) $\mathbb{C}$; (iii) $\{z \in \mathbb{C} \,|\, |z| > 1\}$; (iv) $\{z \in \mathbb{C} \,|\, \text{Im } z \leq -2\}$; (v) $\{i^n | n \in \mathbb{Z}^+\}$; (vi) $\{z \in \mathbb{C} \,|\, 2 < |z - 2| < 4\}$

Proofs:

(i) $\emptyset$ is **open and closed**. The statement that $\emptyset$ is open is vacuously true (since $\emptyset$ has no elements, there is nothing to check). $\emptyset$ is closed because $\mathbb{C} \setminus \emptyset = \mathbb{C}$ is open (proof below). $\square$

(ii) $\mathbb{C}$ is **open and closed**. $\mathbb{C}$ is closed because $\mathbb{C} \setminus \mathbb{C} = \emptyset$ is open (see (i)). To see that $\mathbb{C}$ is open. Let $a \in \mathbb{C}$, and let $D = \{z \in \mathbb{C} \mid |z - a| < 1\}$ be the open disk with center a and radius 1. Since $|a - a| = 0 < 1$, $a \in D$, and since every element of D is a complex number, $D \subseteq \mathbb{C}$. It follows that $\mathbb{C}$ is open. $\square$

(iii) $S = \{z \in \mathbb{C} \mid |z| > 1\}$ is **open**. To see this, let $w \in S$ and let $r = |w| - 1$. We will show that $N_r(w) \subseteq S$ (recall that $N_r(w) = \{z \in \mathbb{C} \mid |z - w| < r\}$). Let $z \in N_r(w)$. Then we have $|z - w| < r = |w| - 1$. So, using the Triangle Inequality, we have

$$|w| = |(w - z) + z| \leq |w - z| + |z| = |z - w| + |z| < |w| - 1 + |z|$$

Thus, $|w| < |w| - 1 + |z|$, and therefore, $|z| > 1$. So, $z \in S$. Since $z \in N_r(w)$ was arbitrary, we have shown that $N_r(w) \subseteq S$. So, S is open. $\square$

$S = \{z \in \mathbb{C} \mid |z| > 1\}$ is **not closed** because $\mathbb{C} \setminus S = \{z \in \mathbb{C} \mid |z| \leq 1\}$ is not open.

(iv) $S = \{z \in \mathbb{C} \mid \text{Im } z \leq -2\}$ is **not open**. To see this, first note that $\text{Im}(-2i) = -2$, and so, $-2i \in S$. If $N_r(-2i)$ is an r-neighborhood of $-2i$, then $\frac{-4+r}{2}i \in N_r(-2i)$ because $\left|\frac{-4+r}{2}i - (-2i)\right| = \left|\left(\frac{-4+r}{2} + \frac{4}{2}\right)i\right| = \left|\frac{r}{2}i\right| = \frac{r}{2} < r$. $\frac{-4+r}{2}i \notin S$ because $\text{Im}\left(\frac{-4+r}{2}i\right) = \frac{-4+r}{2}$, and $\frac{-4+r}{2} > -\frac{4}{2} = -2$. So, for all $r > 0$, $N_r(-2i) \not\subseteq S$, showing that S is not open. $\square$

S is **closed**. To see this, we show that $T = \mathbb{C} \setminus S = \{z \in \mathbb{C} \mid \text{Im } z > -2\}$ is open.

Let $w \in T$ and let $r = 2 + \text{Im } w$. We will show that $N_r(w) \subseteq T$. Let $z \in N_r(w)$. Then we have $\text{Im } w - \text{Im } z = \text{Im}(w - z) \leq |w - z| = |z - w| < r = 2 + \text{Im } w$. So, $-\text{Im } z < 2$, and therefore, $\text{Im } z > -2$. So, $z \in T$. Since $z \in N_r(w)$ was arbitrary, we have shown that $N_r(w) \subseteq T$. So, T is open, and therefore, $S = \mathbb{C} \setminus T$ is closed. $\square$

(v) Note that $S = \{i^n \mid n \in \mathbb{Z}^+\}$ is a finite set consisting of just four complex numbers. Indeed, $S = \{1, -1, i, -i\}$.

S is **not open**. To see this, let $N_r(i)$ be an arbitrary r-neighborhood of i. Then $i + \frac{r}{2} \in N_r(i)$ because $\left|\left(i + \frac{r}{2}\right) - i\right| = \left|\frac{r}{2}\right| = \frac{r}{2} < r$, but $i + \frac{r}{2} \notin S$ because $i + \frac{r}{2}$ is not equal to $1, -1, i$, or $-i$. $\square$

S is **closed**. To see this, we show that $T = \mathbb{C} \setminus S$ is open.

Let $w \in T$ and let $r = \min\{|w - 1|, |w + 1|, |w - i|, |w + i|\}$.

We now show that $N_r(w) \subseteq T$. Since $r \leq |w - 1|$, $1 \notin N_r(w)$ (otherwise, $|w - 1| < r$). Similarly, $-1, i$, and $-i \notin N_r(w)$. So, if $z \in N_r(w)$, then $z \notin S$, and so, $z \in T$.

By Theorem 7.4, $\mathbb{C} \setminus S$ is open. Therefore, S is closed.

(vi) $S = \{z \in \mathbb{C} \mid 2 < |z - 2| < 4\}$ is **open** and **not closed**.

To see that S is open, let $z \in S$ and let $r = \min\{|z - 2| - 2, 4 - |z - 2|\}$. We show that $D_r(z) \subseteq S$. Let $w \in D_r(z)$. Then $|w - z| < r$. So, $|w - z| < 4 - |z - 2|$. Therefore, we have $|w - z| + |z - 2| < 4$, and so, $|w - 2| = |w - z + z - 2| \leq |w - z| + |z - 2| < 4$.

Also, $|w - z| < |z - 2| - 2$. So, we have

$$2 < |z - 2| - |w - z| = |z - w + w - 2| - |w - z| \leq |z - w| + |w - 2| - |w - z|$$
$$= |w - z| + |w - 2| - |w - z| = |w - 2|.$$

So, $2 < |w - 2| < 4$, and therefore, $w \in S$.

Since $z \in S$ was arbitrary, S is open.

To see that S is not closed, we show that $\mathbb{C} \setminus S = \{z \in \mathbb{C} \mid |z - 2| \leq 2 \text{ or } |z - 2| \geq 4\}$ is not open. To see this, first note that $|6 - 2| = |4| = 4 \geq 4$, and so, $6 \in \mathbb{C} \setminus S$. Let $N_r(6)$ be an r-neighborhood of 6 and let $k = \min\{1, \frac{r}{2}\}$. Then we have $6 - k \in N_r(6)$ because $|(6 - k) - 6| = |-k| = k \leq \frac{r}{2} < r$. However, $6 - k \notin \mathbb{C} \setminus S$. To see this, first observe that $|(6 - k) - 2| = |4 - k|$. If $k = 1$, then $|4 - k| = |4 - 1| = |3| = 3$ and it follows that $6 - k \notin \mathbb{C} \setminus S$. If $k = \frac{r}{2}$, then $0 < \frac{r}{2} \leq 1$, so that $-1 \leq -\frac{r}{2} < 0$, and thus, $3 < 4 - \frac{r}{2} < 4$. So, $3 < \left|4 - \frac{r}{2}\right| < 4$ and once again, $6 - k \notin \mathbb{C} \setminus S$. So, $\mathbb{C} \setminus S$ is not open. Therefore, S is not closed. □

8. Prove the following: (i) An arbitrary union of open sets in $\mathbb{C}$ is an open set in $\mathbb{C}$. (ii) A finite intersection of open sets in $\mathbb{C}$ is an open set in $\mathbb{C}$. (iii) An arbitrary intersection of closed sets in $\mathbb{C}$ is a closed set in $\mathbb{C}$. (iv) A finite union of closed sets in $\mathbb{C}$ is a closed set in $\mathbb{C}$. (v) Every open set in $\mathbb{C}$ can be expressed as a union of open disks.

Proofs:

(i) Let X be a set of open subsets of $\mathbb{C}$ and let $z \in \bigcup X$. Then $z \in A$ for some $A \in X$. Since A is open in $\mathbb{C}$, there is an open disk D with $z \in D$ and $D \subseteq A$. By Problem 9 from Lesson 6 (part (i)), we have $A \subseteq \bigcup X$. Since $\subseteq$ is transitive (Theorem 2.3 from Lesson 2), $D \subseteq \bigcup X$. Therefore, $\bigcup X$ is open. □

(ii) Let X be a finite set of open sets in $\mathbb{C}$. If $\bigcap X = \emptyset$, then $\bigcap X$ is open by the proof of Problem 7 (part (i)). Otherwise, let $z \in \bigcap X$. By Theorem 7.4, for each A in X, there is an open disk D_A with center z and radius r_A such that $z \in D_A$ and $D_A \subseteq A$. Let $r = \min\{r_A \mid A \in X\}$ and let D be the open disk with center z and radius r. Since $D = D_A$ for some $A \in X$, $z \in D$. Let $w \in D$ and let $B \in X$. Then $|z - w| < r \leq r_B$. So, $w \in D_B$. Since $B \in X$ was arbitrary, $w \in \bigcap X$. Therefore, $D \subseteq \bigcap X$, and so, $\bigcap X$ is open. □

(iii) Let X be a nonempty set of closed sets in $\mathbb{C}$. Then for each $A \in X$, $\mathbb{C} \setminus A$ is an open set in $\mathbb{C}$. By (i), $\bigcup\{\mathbb{C} \setminus A \mid A \in X\}$ is open in $\mathbb{C}$. Therefore, $\mathbb{C} \setminus \bigcup\{\mathbb{C} \setminus A \mid A \in X\}$ is closed in $\mathbb{C}$. So, it suffices to show that $\bigcap X = \mathbb{C} \setminus \bigcup\{\mathbb{C} \setminus A \mid A \in X\}$. Well, $x \in \bigcap X$ if and only if for all $A \in X$, $x \in A$ if and only if for all $A \in X$, $x \notin \mathbb{C} \setminus A$ if and only if $x \notin \bigcup\{\mathbb{C} \setminus A \mid A \in X\}$ if and only if $x \in \mathbb{C} \setminus \bigcup\{\mathbb{C} \setminus A \mid A \in X\}$. So, $\bigcap X = \mathbb{C} \setminus \bigcup\{\mathbb{C} \setminus A \mid A \in X\}$, completing the proof. □

(iv) Let X be a finite set of closed subsets of $\mathbb{C}$. Then for each $A \in X$, $\mathbb{C} \setminus A$ is an open set in $\mathbb{C}$. By (ii), $\cap\{\mathbb{C} \setminus A \mid A \in X\}$ is open in $\mathbb{C}$. Therefore, $\mathbb{C} \setminus \cap\{\mathbb{C} \setminus A \mid A \in X\}$ is closed in $\mathbb{C}$. So, it suffices to show that $\cup X = \mathbb{C} \setminus \cap\{\mathbb{C} \setminus A \mid A \in X\}$. Well, $x \in \cup X$ if and only if there is an $A \in X$ such that $x \in A$ if and only if there is an $A \in X$ such that $x \notin \mathbb{C} \setminus A$ if and only if $x \notin \cap\{\mathbb{C} \setminus A \mid A \in X\}$ if and only if $x \in \mathbb{C} \setminus \cap\{\mathbb{C} \setminus A \mid A \in X\}$. Therefore, we have $\cup X = \mathbb{C} \setminus \cap\{\mathbb{C} \setminus A \mid A \in X\}$, completing the proof. □

(v) Let X be an open set in $\mathbb{C}$. Since X is open, for each $z \in X$, there is an open disk D_z with $z \in D_z$ and $D_z \subseteq X$. Let $Y = \{D_z \mid z \in X\}$. We will show that $X = \cup Y$.

First, let $z \in X$. Then $z \in D_z$. Since $D_z \in Y$, $z \in \cup Y$. Since z was arbitrary, $X \subseteq \cup Y$.

Now, let $z \in \cup Y$. Then there is $w \in X$ with $z \in D_w$. Since $D_w \subseteq X$, $z \in X$. Since z was arbitrary, $\cup Y \subseteq X$.

Since $X \subseteq \cup Y$ and $\cup Y \subseteq X$, it follows that $X = \cup Y$. □

LEVEL 5

9. A complex number z is an **interior point** of a set S of complex numbers if there is a neighborhood of z that contains only points in S, whereas w is a **boundary point** of S if each neighborhood of w contains at least one point in S and one point not in S. Prove the following: (i) A set of complex numbers is open if and only if each point in S is an interior point of S. (ii) A set of complex numbers is open if and only if it contains none of its boundary points. (iii) A set of complex numbers is closed if and only if it contains all its boundary points.

Proofs:

(i) Let S be a set of complex numbers. Then S is open if and only if for every complex number $z \in S$, there is an open disk D with $z \in D$ and $D \subseteq S$ if and only if for every complex number $z \in S$, there is a neighborhood of z that contains only points in S if and only if every complex number in S is an interior point of S. □

(ii) Suppose that S is an open set of complex numbers and let $z \in S$. By (i), z is an interior point of S. So, there is a neighborhood of z containing only points of S. So, z is **not** a boundary point of S. Since $z \in S$ was arbitrary, S contains none of its boundary points.

We now prove that if S contains none of its boundary points, then S is open by contrapositive. Suppose S is not open. By (i), there is $z \in S$ such that z is **not** an interior point. Let N be a neighborhood of z. Since $z \in S$, N contains a point in S (namely, z). Since z is not an interior point of S, N contains a point not in S. So, z is a boundary point of S. Therefore, S contains at least one of its boundary points. □

(iii) First note that a complex number z is a boundary point of S if and only if z is a boundary point of $\mathbb{C} \setminus S$ (because $z \in S$ if and only if $z \notin \mathbb{C} \setminus S$, and vice versa).

Let S be a set of complex numbers. Then S is closed if and only if $\mathbb{C} \setminus S$ is open if and only if $\mathbb{C} \setminus S$ contains none of its boundary points (by (ii)) if and only if $S = \mathbb{C} \setminus (\mathbb{C} \setminus S)$ contains all its boundary points. □

10. Let $D = \{z \in \mathbb{C} \mid |z| \leq 1\}$ be the closed unit disk and let S be a subset of D that includes the interior of the disk, but is missing at least one point on the bounding circle of the disk. Show that S is not a closed set.

Proof: Let S be a set of complex numbers such that $S \subseteq D$, where $D = \{z \in \mathbb{C} \mid |z| \leq 1\}$, such that S contains $\{z \in \mathbb{C} \mid |z| < 1\}$, but is missing some point w with $|w| = 1$. We will show that w is a boundary point of S. To see this, let N be a neighborhood of w with radius r. w is a point in N that is not in S. We need to find a point in N that is in S.

If $r > 1$, let $z = 0$. Since $0 < 1$, $z \in S$. Also, $|z - w| = |0 - w| = |-w| = |w| = 1 < r$. So, $z \in N$.

If $r \leq 1$, let $z = \frac{2-r}{2} w$. Then $z \in N$ because we have

$$|z - w| = \left|\frac{2-r}{2}w - w\right| = \left|\frac{2-r}{2}w - \frac{2}{2}w\right| = \left|\frac{2-r-2}{2}w\right| = \left|-\frac{r}{2}w\right| = \left|-\frac{r}{2}\right||w| = \frac{r}{2} \cdot 1 = \frac{r}{2} < r.$$

Also, we have $z \in S$ because $|z| = \left|\frac{2-r}{2}w\right| = \left|\frac{2-r}{2}\right||w| = \left|1 - \frac{r}{2}\right|(1) = 1 - \frac{r}{2} < 1$.

So, we have found a boundary point of S that is not in S. By Problem 9, part (iii), S is **not** closed. □

11. Prove that a set of complex numbers is closed if and only if it contains all its accumulation points. (See Problem 6 for the definition of an accumulation point.)

Proof: Suppose that S is a closed set of complex numbers and let a be an accumulation point of S. Assume toward contradiction that $a \notin S$, and let N be a neighborhood of a. Since a is an accumulation point, N contains a point in S. Since $a \notin S$, N contains a point not in S (namely, a). So, a is a boundary point of S. Since S is closed, by Problem 9 (part (iii)), $a \in S$, contradicting our assumption that $a \notin S$. So, we must have $a \in S$. Since a was an arbitrary accumulation point of S, we see that S contains all its accumulation points.

Now, suppose that S contains all its accumulation points, and let a be a boundary point of S. Assume toward contradiction that $a \notin S$. Then each neighborhood of a contains a point in S that is not equal to a. So, each deleted neighborhood of a contains a point in S. So, a is an accumulation point of S, and therefore, by our assumption that S contains all its accumulation points, $a \in S$. This contradicts our assumption that $a \notin S$. So, we must have $a \in S$. Since a was an arbitrary boundary point of S, we see that S contains all its boundary points. By Problem 9 (part (iii)), S is closed. □

12. Prove that a set consisting of finitely many complex numbers is a closed set in $\mathbb{C}$. (Hint: Show that a finite set has no accumulation points.)

Proof: Let S be a set consisting of finitely many points. We will show that S has no accumulation points. Let $a \in \mathbb{C}$ and let $r = \min\{|w - a| \mid w \in S \wedge w \neq a\}$. Suppose toward contradiction that the deleted neighborhood $N_r^{\odot}(a) = \{z \mid 0 < |z - a| < r\}$ contains a complex number in S. Let's call this complex number w. Since $w \in S$ and $w \neq a$, by the definition of r, we have $|w - a| \geq r$. Since $w \in N_r^{\odot}(a)$, we have $|w - a| < r$. So, $r \leq |w - a| < r$, and therefore, $r < r$, a contradiction. Therefore, a is not an accumulation point of S. Since $a \in \mathbb{C}$ was arbitrary, we have shown that S has no accumulation points.

Since S has no accumulation points, the statement "S contains all its accumulation points" is vacuously true. By Problem 11, S is closed. □

Problem Set 8

LEVEL 1

1. Determine if each of the following subsets of $\mathbb{R}^2$ is a subspace of $\mathbb{R}^2$:
   (i) $A = \{(x,y) \mid x + y = 0\}$; (ii) $B = \{(x,y) \mid xy = 0\}$; (iii) $C = \{(x,y) \mid 2x = 3y\}$;
   (iv) $D = \{(x,y) \mid x \in \mathbb{Q}\}$

Solutions:

(i) Since $0 + 0 = 0$, $(0,0) \in A$.

   Let $(x, y), (z, w) \in A$. Then $x + y = 0$ and $z + w = 0$. Therefore,
   $$(x + z) + (y + w) = (x + y) + (z + w) = 0 + 0 = 0.$$
   So, $(x, y) + (z, w) = (x + z, y + w) \in A$.

   Let $(x, y) \in A$ and $k \in \mathbb{R}$. Then $x + y = 0$. So, $kx + ky = k(x + y) = k \cdot 0 = 0$ (by part (iii) of Problem 4 below).

   So, $k(x, y) = (kx, ky) \in A$.

   By Theorem 8.1, A is a subspace of $\mathbb{R}^2$.

(ii) Since $0 \cdot 1 = 0$, we have $(0, 1) \in B$. Since $1 \cdot 0 = 0$, we have $(1, 0) \in B$. Adding these two vectors gives us $(1, 0) + (0, 1) = (1, 1)$. However, $1 \cdot 1 = 1 \neq 0$, and so, $(1, 1) \notin B$. So, B is not closed under addition. Therefore, B is **not** a subspace of $\mathbb{R}^2$.

(iii) Since $2 \cdot 0 = 0$ and $3 \cdot 0 = 0$, $2 \cdot 0 = 3 \cdot 0$. Therefore, $(0, 0) \in C$.

   Let $(x, y), (z, w) \in C$. Then $2x = 3y$ and $2z = 3w$. Therefore,
   $$2(x + z) = 2x + 2z = 3y + 3w = 3(y + w).$$
   So, $(x, y) + (z, w) = (x + z, y + w) \in C$.

   Let $(x, y) \in C$ and $k \in \mathbb{R}$. Then $2x = 3y$. So, $2(kx) = k(2x) = k(3y) = 3(ky)$.

   So, $k(x, y) = (kx, ky) \in C$.

   By Theorem 8.1, A is a subspace of $\mathbb{R}^2$.

(iv) Since $1 \in \mathbb{Q}$, $(1, 0) \in D$. Now, $\sqrt{2}(1, 0) = (\sqrt{2}, 0) \notin D$ because $\sqrt{2} \notin \mathbb{Q}$. So, D is not closed under scalar multiplication. Therefore, D is **not** a subspace of $\mathbb{R}^2$.

2. For each of the following, determine if the given pair of vectors v and w are linearly independent or linearly dependent in the given vector space V: (i) $V = \mathbb{Q}^4$, $v = (3, 2, 2, -1)$, $w = \left(-1, -\frac{2}{3}, -\frac{2}{3}, -\frac{1}{3}\right)$; (ii) $V = \mathbb{R}^3$, $v = (1, \sqrt{2}, 1)$, $w = (\sqrt{2}, 2, \sqrt{2})$; (iii) $V = \mathbb{C}^5$, $v = (1, i, 2-i, 0, 3i)$, $w = (-i, 1, -1 - 2i, 0, 3)$; (iv) $V = M_{22}^{\mathbb{Q}}$, $v = \begin{bmatrix} a & b \\ a & 3b \\ 2 & \end{bmatrix}$, $w = \begin{bmatrix} 1 & \frac{b}{a} \\ \frac{1}{2} & 3 \end{bmatrix}$

   $(a \neq 0, a \neq b)$; (v) $V = \{ax^2 + bx + c \mid a, b, c \in \mathbb{R}\}$, $v = x, w = x^2$

Solutions:

(i) $-3w = -3\left(-1, -\frac{2}{3}, -\frac{2}{3}, -\frac{1}{3}\right) = (3, 2, 2, 1)$. Since $-3(-1) = 3$, but $-3\left(-\frac{1}{3}\right) \neq -1$, v and w are **not** scalar multiples of each other. Therefore, v and w are **linearly independent**.

(ii) $\sqrt{2}v = \sqrt{2}(1, \sqrt{2}, 1) = (\sqrt{2}, 2, \sqrt{2}) = w$. So, v and w **are** scalar multiples of each other. Therefore, v and w are **linearly dependent**.

(iii) $-iv = -i(1, i, 2-i, 0, 3i) = (-i, 1, -1-2i, 0, 3) = w$. So, v and w **are** scalar multiples of each other. Therefore, v and w are **linearly dependent**.

(iv) $aw = a\begin{bmatrix} 1 & \frac{b}{a} \\ \frac{1}{2} & 3 \end{bmatrix} = \begin{bmatrix} a & b \\ \frac{a}{2} & 3a \end{bmatrix}$. Since $a \cdot 1 = a$, but $a \cdot 3 = 3a \neq 3b$, v and w are **not** scalar multiples of each other. Therefore, v and w are **linearly independent**.

(v) If $k \in \mathbb{R}$, then $kx \neq x^2$. So, x and x^2 are **not** scalar multiples of each other. Therefore, v and w are **linearly independent**.

LEVEL 2

3. Let $\mathbb{F}$ be a field. Prove that $\mathbb{F}^n$ is a vector space over $\mathbb{F}$.

Proof: We first prove that $(\mathbb{F}^n, +)$ is a commutative group.

(Closure) Let $(a_1, a_2, \ldots, a_n), (b_1, b_2, \ldots, b_n) \in \mathbb{F}^n$. Then $a_1, a_2, \ldots, a_n, b_1, b_2, \ldots, b_n \in \mathbb{F}$. By definition, $(a_1, a_2, \ldots, a_n) + (b_1, b_2, \ldots, b_n) = (a_1 + b_1, a_2 + b_2, \ldots, a_n + b_n)$. Since $\mathbb{F}$ is closed under addition, $a_1 + b_1, a_2 + b_2, \ldots, a_n + b_n \in \mathbb{F}$. Therefore, $(a_1, a_2, \ldots, a_n) + (b_1, b_2, \ldots, b_n) \in \mathbb{F}^n$.

(Associativity) Let $(a_1, a_2, \ldots, a_n), (b_1, b_2, \ldots, b_n), (c_1, c_2, \ldots, c_n) \in \mathbb{F}^n$. Since addition is associative in $\mathbb{F}$, we have

$$[(a_1, a_2, \ldots, a_n) + (b_1, b_2, \ldots, b_n)] + (c_1, c_2, \ldots, c_n) = (a_1 + b_1, a_2 + b_2, \ldots, a_n + b_n) + (c_1, c_2, \ldots, c_n)$$
$$= ((a_1 + b_1) + c_1, (a_2 + b_2) + c_2, \ldots, (a_n + b_n) + c_n)$$
$$= (a_1 + (b_1 + c_1), a_2 + (b_2 + c_2), \ldots, a_n + (b_n + c_n))$$
$$= (a_1, a_2, \ldots, a_n) + (b_1 + c_1, b_2 + c_2, \ldots, b_n + c_n)$$
$$= (a_1, a_2, \ldots, a_n) + [(b_1, b_2, \ldots, b_n) + (c_1, c_2, \ldots, c_n)].$$

(Commutativity) Let $(a_1, a_2, \ldots, a_n), (b_1, b_2, \ldots, b_n) \in \mathbb{F}^n$. Since addition is commutative in $\mathbb{R}$, we have
$$(a_1, a_2, \ldots, a_n) + (b_1, b_2, \ldots, b_n) = (a_1 + b_1, a_2 + b_2, \ldots, a_n + b_n) = (b_1 + a_1, b_2 + a_2, \ldots, b_n + a_n)$$
$$= (b_1, b_2, \ldots, b_n) + (a_1, a_2, \ldots, a_n).$$

(Identity) We show that $(0, 0, \ldots, 0)$ is an additive identity for $\mathbb{F}^n$. Let $(a_1, a_2, \ldots, a_n) \in \mathbb{F}^n$. Since 0 is an additive identity for $\mathbb{R}$, we have

$$(0, 0, \ldots, 0) + (a_1, a_2, \ldots, a_n) = (0 + a_1, 0 + a_2, \ldots, 0 + a_n) = (a_1, a_2, \ldots, a_n).$$
$$(a_1, a_2, \ldots, a_n) + (0, 0, \ldots, 0) = (a_1 + 0, a_2 + 0, \ldots, a_n + 0) = (a_1, a_2, \ldots, a_n).$$

(Inverse) Let $(a_1, a_2, ..., a_n) \in \mathbb{F}^n$. Then $a_1, a_2, ..., a_n \in \mathbb{F}$. Since $\mathbb{F}$ has the additive inverse property, $-a_1, -a_2, ..., -a_n \in \mathbb{F}$. So, $(-a_1, -a_2, ..., -a_n) \in \mathbb{F}^n$ and

$$(a_1, a_2, ..., a_n) + (-a_1, -a_2, ..., -a_n) = (a_1 - a_1, a_2 - a_2, ..., a_n - a_n) = (0, 0, ..., 0).$$
$$(-a_1, -a_2, ..., -a_n) + (a_1, a_2, ..., a_n) = (-a_1 + a_1, -a_2 + a_2, ..., -a_n + a_n) = (0, 0, ..., 0).$$

Now, let's prove that $\mathbb{F}^n$ has the remaining vector space properties.

(Closure under scalar multiplication) Let $k \in \mathbb{F}$ and let $(a_1, a_2, ..., a_n) \in \mathbb{F}^n$. Then $a_1, a_2, ..., a_n \in \mathbb{F}$. By definition, $k(a_1, a_2, ..., a_n) = (ka_1, ka_2, ..., ka_n)$. Since $\mathbb{F}$ is closed under multiplication, $ka_1, ka_2, ..., ka_n \in \mathbb{F}$. Therefore, $k(a_1, a_2, ..., a_n) \in \mathbb{F}^n$.

(Scalar multiplication identity) Let 1 be the multiplicative identity of $\mathbb{F}$ and let $(a_1, a_2, ..., a_n) \in \mathbb{F}^n$. Then $1(a_1, a_2, ..., a_n) = (1a_1, 1a_2, ..., 1a_n) = (a_1, a_2, ..., a_n)$.

(Associativity of scalar multiplication) Let $j, k \in \mathbb{F}$ and $(a_1, a_2, ..., a_n) \in \mathbb{F}^n$. Then since multiplication is associative in $\mathbb{F}$, we have

$$(jk)(a_1, a_2, ..., a_n) = ((jk)a_1, (jk)a_2, ..., (jk)a_n) = (j(ka_1), j(ka_2), ..., j(ka_n))$$
$$= j(ka_1, ka_2, ..., ka_n) = j(k(a_1, a_2, ..., a_n)).$$

(Distributivity of 1 scalar over 2 vectors) Let $k \in \mathbb{F}$ and $(a_1, a_2, ..., a_n), (b_1, b_2, ..., b_n) \in \mathbb{F}^n$. Since multiplication is distributive over addition in $\mathbb{F}$, we have

$$k((a_1, a_2, ..., a_n) + (b_1, b_2, ..., b_n)) = k((a_1 + b_1, a_2 + b_2, ..., a_n + b_n))$$
$$= (k(a_1 + b_1), k(a_2 + b_2), ..., k(a_n + b_n)) = ((ka_1 + kb_1), (ka_2 + kb_2), ..., (ka_n + kb_n))$$
$$= (ka_1, ka_2, ..., ka_n) + (kb_1, kb_2, ..., kb_n) = k(a_1, a_2, ..., a_n) + k(b_1, b_2, ..., b_n).$$

(Distributivity of 2 scalars over 1 vector) Let $j, k \in \mathbb{F}$ and $(a_1, a_2, ..., a_n) \in \mathbb{F}^n$. Since multiplication is distributive over addition in $\mathbb{F}$, we have

$$(j + k)(a_1, a_2, ..., a_n) = ((j + k)a_1, (j + k)a_2, ..., (j + k)a_n)$$
$$= (ja_1 + ka_1, ja_2 + ka_2, ..., ja_n + ka_n) = (ja_1, ja_2, ..., ja_n) + (ka_1, ka_2, ..., ka_n)$$
$$= j(a_1, a_2, ..., a_n) + k(a_1, a_2, ..., a_n).$$

4. Let V be a vector space over $\mathbb{F}$. Prove each of the following: (i) For every $v \in V$, $-(-v) = v$; (ii) For every $v \in V$, $0v = 0$; (iii) For every $k \in \mathbb{F}$, $k \cdot 0 = 0$; (iv) For every $v \in V$, $-1v = -v$

Proofs:

(i) Since $-v$ is the additive inverse of v, we have $v + (-v) = -v + v = 0$. But this equation also says that v is the additive inverse of $-v$. So, $-(-v) = v$. □

(ii) Let $v \in V$. Then $0v = (0 + 0)v = 0v + 0v$. So, we have

$$0 = -0v + 0v = -0v + (0v + 0v) = (-0v + 0v) + 0v = 0 + 0v = 0v.$$ □

(iii) Let $k \in \mathbb{F}$. Then $k \cdot 0 = k(0 + 0) = k \cdot 0 + k \cdot 0$. So, we have

$$0 = -k \cdot 0 + k \cdot 0 = -k \cdot 0 + (k \cdot 0 + k \cdot 0) = (-k \cdot 0 + k \cdot 0) + k \cdot 0 = 0 + k \cdot 0 = k \cdot 0.$$ □

(iv) Let $v \in V$. Then we have $v + (-1v) = 1v + (-1v) = (1 + (-1))v = 0v = 0$ by (ii) and we have $-1v + v = -1v + 1v = (-1 + 1)v = 0v = 0$ again by (ii). So, $-1v = -v$. □

LEVEL 3

5. Let V be a vector space over a field $\mathbb{F}$ and let X be a set of subspaces of V. Prove that $\bigcap X$ is a subspace of V.

Proof: Let V be a vector space over a field $\mathbb{F}$ and let X be a set of subspaces of V. For each $U \in X$, $0 \in U$ because $U \le V$. So, $0 \in \bigcap X$. Let $v, w \in \bigcap X$. For each $U \in X$, $v, w \in U$, and so, $v + w \in U$ because $U \le V$. Therefore, $v + w \in \bigcap X$. Let $v \in \bigcap X$ and $k \in \mathbb{F}$. For each $U \in X$, $v \in U$, and so, $kv \in U$ because $U \le V$. Therefore, $kv \in \bigcap X$. By Theorem 8.1, $\bigcap X \le V$. □

6. Prove that a finite set with at least two vectors is linearly dependent if and only if one of the vectors in the set can be written as a linear combination of the other vectors in the set.

Proof: Suppose that $S = \{v_1, v_2, \ldots, v_n\}$ is a linearly dependent set with at least two elements. Then there are weights $c_1, c_2, \ldots, c_n$ not all 0 such that $c_1 v_1 + c_2 v_2 + \cdots + c_n v_n = 0$. Without loss of generality, assume that $c_1 \ne 0$. We have $c_1 v_1 = -c_2 v_2 - \cdots - c_n v_n$, and so, $v_1 = -\frac{c_2}{c_1} v_2 - \cdots - \frac{c_n}{c_1} v_n$. So, v_1 can be written as a linear combination of the other vectors in S.

Now, suppose that one of the vectors in S can be written as a linear combination of the other vectors in the set. Without loss of generality, assume that $v_1 = c_2 v_2 + \cdots + c_n v_n$. Then we have

$$v_1 - c_2 v_2 - \cdots - c_n v_n = 0.$$

Since the weight of v_1 is 1, this is a nontrivial dependence relation. This shows that S is a linearly dependent set. □

LEVEL 4

7. Let U and W be subspaces of a vector space V. Determine necessary and sufficient conditions for $U \cup W$ to be a subspace of V.

Theorem: Let U and W be subspaces of a vector space V. Then $U \cup W$ is a subspace of V if and only if $U \subseteq W$ or $W \subseteq U$.

Proof: Let U and W be subspaces of a vector space V. If $U \subseteq W$, then $U \cup W = W$, and so, $U \cup W$ is a subspace of V. Similarly, if $W \subseteq U$, then $U \cup W = U$, and so, $U \cup W$ is a subspace of V.

Suppose that $U \nsubseteq W$ and $W \nsubseteq U$. Let $x \in U \setminus W$ and $y \in W \setminus U$. Suppose that $x + y \in U$. We have $-x \in U$ because U is a subspace of V. So, $y = (-x + x) + y = -x + (x + y) \in U$, contradicting $y \in W \setminus U$. So, $x + y \notin U$. A similar argument shows that $x + y \notin W$. So, $x + y \notin U \cup W$. It follows that $U \cup W$ is not closed under addition, and therefore, $U \cup W$ is **not** a subspace of V. □

Note: The conditional statement $p \to q$ can be read "q is necessary for p" or "p is sufficient for q." Furthermore, $p \leftrightarrow q$ can be read "p is necessary and sufficient for q" (as well as "q is necessary and sufficient for p").

So, when we are asked to determine necessary and sufficient conditions for a statement p to be true, we are being asked to find a statement q that is logically equivalent to the statement p.

Usually if we are being asked for necessary and sufficient conditions, the hope is that we will come up with an equivalent statement that is easier to understand and/or visualize than the given statement.

8. Give an example of vector spaces U and V with $U \subseteq V$ such that U is closed under scalar multiplication, but U is not a subspace of V.

Solution: Let $V = \mathbb{R}^2$ and $U = \{(x,y) \mid x = 0 \text{ or } y = 0 \text{ (or both)}\}$. Let $(x,y) \in U$ and $k \in \mathbb{R}$. Then $k(x,y) = (kx, ky)$. If $x = 0$, then $kx = 0$. If $y = 0$, then $ky = 0$. So, $k(x,y) \in U$. So, U is closed under scalar multiplication. Now, $(0,1)$ and $(1,0)$ are in U, but $(1,1) = (0,1) + (1,0) \notin U$. So, $U \not\leq V$. □

LEVEL 5

9. Let S be a set of two or more linearly dependent vectors in a vector space V. Prove that there is a vector v in the set so that span $S = \text{span } S \setminus \{v\}$.

Proof: Let $S = \{v_1, v_2, \ldots, v_n\}$ be a set of two or more linearly dependent vectors in V. By Problem 6, one of the vectors in the set can be written as a linear combination of the other vectors in the set. Without loss of generality, assume that v_n can be written as a linear combination of the other vectors in the set, say $v_n = k_1 v_1 + k_2 v_2 + \cdots + k_{n-1} v_{n-1}$. We show that span $S = \text{span } S \setminus \{v_n\}$. Let $v \in \text{span } S$. Then there are weights $c_1, c_2, \ldots, c_n$ with $v = c_1 v_1 + c_2 v_2 + \cdots + c_n v_n$. So, we have

$$v = c_1 v_1 + c_2 v_2 + \cdots + c_n v_n = c_1 v_1 + c_2 v_2 + \cdots + c_n(k_1 v_1 + k_2 v_2 + \cdots + k_{n-1} v_{n-1})$$
$$= (c_1 + c_n k_1)v_1 + (c_2 + c_n k_2)v_2 + \cdots + (c_{n-1} + c_n k_{n-1})v_{n-1} \in \text{span } S \setminus \{v_n\}.$$

So, span $S \subseteq \text{span } S \setminus \{v_n\}$. Since it is clear that span $S \setminus \{v_n\} \subseteq \text{span } S$, span $S = \text{span } S \setminus \{v_n\}$. □

10. Prove that a finite set of vectors S in a vector space V is a basis of V if and only if every vector in V can be written uniquely as a linear combination of the vectors in S.

Proof: Suppose that $S = \{v_1, v_2, \ldots, v_n\}$ is a basis of V. Then $\text{span}\{v_1, v_2, \ldots, v_n\} = V$. So, if $v \in V$, then v can be written as a linear combination of the vectors in S. Suppose there are weights $c_1, c_2, \ldots, c_n$ and $d_1, d_2, \ldots, d_n$ such that $v = c_1 v_1 + c_2 v_2 + \cdots + c_n v_n$ and $v = d_1 v_1 d c_2 v_2 + \cdots + d_n v_n$. Then we have $c_1 v_1 + c_2 v_2 + \cdots + c_n v_n = d_1 v_1 d c_2 v_2 + \cdots + d_n v_n$, and so,

$$(c_1 - d_1)v_1 + (c_2 - d_2)v_2 + \cdots + (c_n - d_n)v_n = 0.$$

Since S is a linearly independent set of vectors, $c_1 - d_1 = 0, c_2 - d_2 = 0, \ldots, c_n - d_n = 0$, and therefore, $c_1 = d_1, c_2 = d_2, \ldots, c_n = d_n$. So, the expression of v as a linear combination of the vectors in S is unique.

Now, suppose that each vector in V can be written uniquely as a linear combination of the vectors in S. Since each vector in V can be written as a linear combination of the vectors in S, we have that $\text{span}\{v_1, v_2, \ldots, v_n\} = V$. Since $0v_1 + 0v_2 + \cdots + 0v_n = 0$, by the uniqueness condition, the only way $c_1 v_1 + c_2 v_2 + \cdots + c_n v_n = 0$ could be true is if all weights are 0. So, S is linearly independent, and therefore, S is a basis of V.

11. Let $S = \{v_1, v_2, \ldots, v_m\}$ be a set of linearly independent vectors in a vector space V and let $T = \{w_1, w_2, \ldots, w_n\}$ be a set of vectors in V such that span $T = V$. Prove that $m \leq n$.

Proof: If $V = \{0\}$ or S consists of just one vector, then there is nothing to prove. So, let's assume that $V \neq \{0\}$ and S has at least two vectors. Note that since $V \neq \{0\}$, T has at least one vector.

Let $T_0 = T = \{w_1, w_2, \ldots, w_n\}$. Since span $T_0 = V$, v_1 can be written as a linear combination of the vectors in T_0. By Problem 6, $\{w_1, w_2, \ldots, w_n, v_1\}$ is linearly dependent. Let $c_1, c_2, \ldots, c_n, d$ be weights, not all of which are 0, such that $c_1 w_1 + c_2 w_2 + \cdots + c_n w_n + d v_1 = 0$. We claim that for some $i = 1, 2, \ldots, n$, $c_i \neq 0$. If $d = 0$, then since one of the weights must be nonzero, some c_i must be nonzero. Suppose $d \neq 0$. If every $c_i = 0$, then $dv_1 = 0$. Since $d \neq 0$, $v_1 = 0$, contradicting the linear independence of S. In both cases, we must have $c_i \neq 0$ for some i. Without loss of generality, assume that $c_1 \neq 0$. Then $w_1 = -\frac{c_2}{c_1} w_2 - \cdots - \frac{c_n}{c_1} w_n - \frac{d}{c_1} v_1$. Let $T_1 = \{w_2, \ldots, w_n, v_1\}$. By the proof of Problem 9, we have span $\{w_1, w_2, \ldots, w_n, v_1\} \subseteq$ span T_1. So, span $T_1 = V$.

At this point, note that if T had just one vector, then $T_1 = \{v_1\}$. Since span $T_1 = V$, v_2 would be a scalar multiple of v_1, contradicting the linear independence of S. So, T has at least two vectors. If S has only two vectors, then we are done. Otherwise, we continue as follows.

Since span $T_1 = V$, $\{w_2, \ldots, w_n, v_1, v_2\}$ is linearly dependent. Let $c_2, \ldots, c_n, d_1, d_2$ be weights, not all of which are 0, such that $c_2 w_2 + \cdots + c_n w_n + d_1 v_1 + d_2 v_2 = 0$. We claim that for some $i = 2, \ldots, n$, $c_i \neq 0$. If $d_1 = 0$ or $d_2 = 0$, we can use the same argument in the last paragraph to show that some c_i must be nonzero. Suppose $d_1 \neq 0$ and $d_2 \neq 0$. If every $c_i = 0$, then $d_1 v_1 + d_2 v_2 = 0$, contradicting the linear independence of S. In both cases, we must have $c_i \neq 0$ for some i. Without loss of generality, assume that $c_2 \neq 0$. Then $w_2 = -\frac{c_3}{c_2} w_3 - \cdots - \frac{c_n}{c_2} w_n - \frac{d_1}{c_2} v_1 - \frac{d_2}{c_2} v_2$. Let $T_2 = \{w_3, \ldots, w_n, v_1, v_2\}$. By the proof of Problem 9, we have span $\{w_2, \ldots, w_n, v_1, v_2\} \subseteq$ span T_2. So, span $T_2 = V$.

Observe that if T had just two vectors, then $T_2 = \{v_1, v_2\}$. Since span $T_2 = V$, v_3 could be written as a linear combination of v_1 and v_2, contradicting the linear independence of S. So, T has at least three vectors. If S has only three vectors, then we are done. Otherwise, we continue in the same way.

Assuming $T_{j-1} = \{w_j, \ldots, w_n, v_1, v_2, \ldots v_{j-1}\}$ and span $T_{j-1} = V$, we have $\{w_j, \ldots, w_n, v_1, v_2, \ldots v_{j-1}, v_j\}$ linearly dependent. Once again, reindexing the w_i's if necessary, and letting $T_j = \{w_{j+1}, \ldots, w_n, v_1, v_2, \ldots v_{j-1}, v_j\}$, by an argument just like that given in the first paragraph, we can show that span $\{w_j, \ldots, w_n, v_1, v_2, \ldots, v_j\} \subseteq$ span T_j. So, span $T_j = V$.

If $j < m$ and T had just j vectors, then $T_j = \{v_1, v_2, \ldots, v_j\}$. Since span $T_j = V$, v_{j+1} could be written as a linear combination of the vectors $v_1, v_2, \ldots, v_j$, contradicting the linear independence of S. So, T has at least $j + 1$ vectors. If $j = m$, we have shown that $m \leq n$. Otherwise, we continue in the same way. This procedure terminates in m steps. □

12. Let B be a basis of a vector space V with n vectors. Prove that any other basis of V also has n vectors.

Proof: Let B be a basis of V with n vectors. Let B' be another basis of V with m vectors. Since B' is a basis of V, B' is a linearly independent set of vectors in V. Since B is a basis of V, span $B = V$. By Problem 11, $m \leq n$. Similarly, we have that B is a linearly independent set of vectors in V and span $B' = V$. So, $n \leq m$. Since $m \leq n$ and $n \leq m$, we have $m = n$. So, B' has n vectors. □

Problem Set 9

LEVEL 1

1. Let ϕ be the following statement: $(p \wedge \neg q) \leftrightarrow \neg[p \vee (\neg r \rightarrow q)]$. (i) The statement ϕ is abbreviated. Write ϕ in its unabbreviated form. (ii) Write down all the substatements of ϕ in both abbreviated and unabbreviated form.

Solutions:

(i) $\Big((p \wedge (\neg q)) \leftrightarrow (\neg[p \vee ((\neg r) \rightarrow q)])\Big)$

(ii) Abbreviated forms: $p, q, r, \neg q, \neg r, p \wedge \neg q, \neg r \rightarrow q, p \vee (\neg r \rightarrow q), \neg[p \vee (\neg r \rightarrow q)]$

Unabbreviated forms: $p, q, r, (\neg q), (\neg r), (p \wedge (\neg q)), ((\neg r) \rightarrow q), \big(p \vee ((\neg r) \rightarrow q)\big), \big(\neg[p \vee ((\neg r) \rightarrow q)]\big)$

2. Verify all the logical equivalences given in List 9.1.

Solutions:

1. **Law of double negation:** $p \equiv \neg(\neg p)$: This was done in Example 9.3.

2. **De Morgan's laws:** $\neg(p \wedge q) \equiv \neg p \vee \neg q$: This was done in Example 9.4

    $\neg(p \vee q) \equiv \neg p \wedge \neg q$: Let $\phi = \neg(p \vee q)$ and let $\psi = \neg p \wedge \neg q$. If $p \equiv T$ or $q \equiv T$, then $\phi \equiv \neg T \equiv F$ and $\psi \equiv F$ (because $\neg p \equiv F$ or $\neg q \equiv F$). If $p \equiv F$ and $q \equiv F$, then $\phi \equiv \neg F \equiv T$ and $\psi \equiv T \wedge T \equiv T$. So, all four possible truth assignments of p and q lead to the same truth value for ϕ and ψ. It follows that $\phi \equiv \psi$.

3. **Commutative laws:** $p \wedge q \equiv q \wedge p, p \vee q \equiv q \vee p$: Look at the truth tables.

4. **Associative laws:** $(p \wedge q) \wedge r \equiv p \wedge (q \wedge r), (p \vee q) \vee r \equiv p \vee (q \vee r)$: Draw truth tables.

5. **Distributive laws:** $p \wedge (q \vee r) \equiv (p \wedge q) \vee (p \wedge r), p \vee (q \wedge r) \equiv (p \vee q) \wedge (p \vee r)$: Draw truth tables.

6. **Identity laws:** $p \wedge T \equiv p$: Let $\phi = p \wedge T$ and let $\psi = p$. If $p \equiv T$, then $\phi \equiv T \wedge T \equiv T$ and $\psi \equiv T$. If $p \equiv F$, then $\phi \equiv F \wedge T \equiv F$ and $\psi \equiv F$. So, both possible truth assignments of p lead to the same truth value for ϕ and ψ. It follows that $\phi \equiv \psi$.

    $p \wedge F \equiv F$: Let $\phi = p \wedge F$ and let $\psi = F$. If $p \equiv T$, then $\phi \equiv T \wedge F \equiv F$ and $\psi \equiv F$. If $p \equiv F$, then $\phi \equiv F \wedge F \equiv F$ and $\psi \equiv F$. So, both possible truth assignments of p lead to the same truth value for ϕ and ψ. It follows that $\phi \equiv \psi$.

    $p \vee T \equiv T$: Let $\phi = p \vee T$ and let $\psi = T$. If $p \equiv T$, then $\phi \equiv T \vee T \equiv T$ and $\psi \equiv T$. If $p \equiv F$, then $\phi \equiv F \vee T \equiv T$ and $\psi \equiv T$. So, both possible truth assignments of p lead to the same truth value for ϕ and ψ. It follows that $\phi \equiv \psi$.

    $p \vee F \equiv p$: Let $\phi = p \vee F$ and let $\psi = p$. If $p \equiv T$, then $\phi \equiv T \vee F \equiv T$ and $\psi \equiv T$. If $p \equiv F$, then $\phi \equiv F \vee F \equiv F$ and $\psi \equiv F$. So, both possible truth assignments of p lead to the same truth value for ϕ and ψ. It follows that $\phi \equiv \psi$.

7. **Negation laws:** $p \land \neg p \equiv F$: Let $\phi = p \land \neg p$ and let $\psi = F$. If $p \equiv T$, then $\phi \equiv T \land F \equiv F$ and $\psi \equiv F$. If $p \equiv F$, then $\phi \equiv F \land T \equiv F$ and $\psi \equiv F$. So, both possible truth assignments of p lead to the same truth value for ϕ and ψ. It follows that $\phi \equiv \psi$.

   $p \lor \neg p \equiv T$: Let $\phi = p \lor \neg p$ and let $\psi = T$. If $p \equiv T$, then $\phi \equiv T \lor F \equiv T$ and $\psi \equiv T$. If $p \equiv F$, then $\phi \equiv F \lor T \equiv T$ and $\psi \equiv T$. So, both possible truth assignments of p lead to the same truth value for ϕ and ψ. It follows that $\phi \equiv \psi$.

8. **Redundancy laws:** $p \land p \equiv p$: Let $\phi = p \land p$ and let $\psi = p$. If $p \equiv T$, then $\phi \equiv T \land T \equiv T$ and $\psi \equiv T$. If $p \equiv F$, then $\phi \equiv F \land F \equiv F$ and $\psi \equiv F$. So, both possible truth assignments of p lead to the same truth value for ϕ and ψ. It follows that $\phi \equiv \psi$.

   $p \lor p \equiv p$: Let $\phi = p \lor p$ and let $\psi = p$. If $p \equiv T$, then $\phi \equiv T \lor T \equiv T$ and $\psi \equiv T$. If $p \equiv F$, then $\phi \equiv F \lor F \equiv F$ and $\psi \equiv F$. So, both possible truth assignments of p lead to the same truth value for ϕ and ψ. It follows that $\phi \equiv \psi$.

   Absorption laws: $(p \lor q) \land p \equiv p$: Let $\phi = (p \lor q) \land p$ and let $\psi = p$. If $p \equiv T$, then $p \lor q \equiv T \lor q \equiv T$. So, $\phi \equiv T \land T \equiv T$. Also, $\psi \equiv T$. If $p \equiv F$, then $\phi \equiv (p \lor q) \land F \equiv F$ and $\psi \equiv F$. So, all four possible truth assignments of p and q lead to the same truth value for ϕ and ψ. It follows that $\phi \equiv \psi$.

   $(p \land q) \lor p \equiv p$: Let $\phi = (p \land q) \lor p$ and let $\psi = p$. If $p \equiv T$, then $\phi \equiv (p \land q) \lor T \equiv T$ and $\psi \equiv T$. If $p \equiv F$, then $p \land q \equiv F \land q \equiv F$. So, $\phi \equiv F \lor F \equiv F$. Also, $\psi \equiv F$. So, all four possible truth assignments of p and q lead to the same truth value for ϕ and ψ. It follows that $\phi \equiv \psi$.

9. **Law of the conditional:** $p \to q \equiv \neg p \lor q$: Let $\phi = p \to q$ and let $\psi = \neg p \lor q$. If $p \equiv F$, then $\phi \equiv F \to q \equiv T$ and $\psi \equiv T \lor q \equiv T$. If $q \equiv T$, then $\phi \equiv p \to T \equiv T$ and $\psi \equiv \neg p \lor T \equiv T$. Finally, if $p \equiv T$ and $q \equiv F$, then $\phi \equiv T \to F \equiv F$ and $\psi \equiv F \lor F \equiv F$. So, all four possible truth assignments of p and q lead to the same truth value for ϕ and ψ. It follows that $\phi \equiv \psi$.

10. **Law of the contrapositive:** $p \to q \equiv \neg q \to \neg p$: Let $\phi = p \to q$ and let $\psi = \neg q \to \neg p$. If $p \equiv F$, then $\phi \equiv F \to q \equiv T$ and $\psi \equiv \neg q \to T \equiv T$. If $q \equiv T$, then $\phi \equiv p \to T \equiv T$ and $\psi \equiv F \to \neg p \equiv T$. Finally, if $p \equiv T$ and $q \equiv F$, then $\phi \equiv T \to F \equiv F$ and $\psi \equiv T \to F \equiv F$. So, all four possible truth assignments of p and q lead to the same truth value for ϕ and ψ. It follows that $\phi \equiv \psi$.

11. **Law of the biconditional:** $p \leftrightarrow q \equiv (p \to q) \land (q \to p)$: Let $\phi = p \leftrightarrow q$ and let $\psi = (p \to q) \land (q \to p)$. If $p \equiv T$, then $\phi \equiv T \leftrightarrow q \equiv q$, $\psi \equiv (T \to q) \land (q \to T) \equiv q \land T \equiv q$. If $p \equiv F$, $\phi \equiv F \leftrightarrow q \equiv \neg q$, $\psi \equiv (F \to q) \land (q \to F) \equiv T \land \neg q \equiv \neg q$. So, all four possible truth assignments of p and q lead to the same truth value for ϕ and ψ. It follows that $\phi \equiv \psi$.

LEVEL 2

3. Let ϕ, ψ, and τ be statements. Prove that $\phi \vdash \psi$ and $\psi \vdash \tau$ implies $\phi \vdash \tau$.

Proof: Let ϕ, ψ, and τ be statements with $\phi \vdash \psi$ and $\psi \vdash \tau$. Let a be a truth assignment that makes ϕ true. Since $\phi \vdash \psi$, a makes ψ true. Since $\psi \vdash \tau$, a makes τ true. Since a was an arbitrary truth assignment that makes ϕ true, $\phi \vdash \tau$. □

Notes: (1) Recall that the symbol $\vdash$ is pronounced "tautologically implies."

(2) If a truth assignment a makes a statement ϕ true, we say that a **satisfies** ϕ. If a makes ϕ false, we say that a **does not satisfy** ϕ.

> 4. Let ϕ and ψ be statements. Prove that $\phi \vdash \psi$ if and only if $\phi \to \psi$ is a tautology.

Proof: Let ϕ and ψ be statements and assume that $\phi \vdash \psi$. Let a be a truth assignment of the propositional variables appearing in ϕ or ψ or both. If a satisfies ϕ, then a satisfies ψ (because $\phi \vdash \psi$). It follows that $\phi \to \psi \equiv T \to T \equiv T$. If a does not satisfy ϕ, then $\phi \to \psi \equiv F \to \psi \equiv T$. So, we have shown that every truth assignment makes $\phi \to \psi$ true. Therefore, $\phi \to \psi$ is a tautology.

Conversely, assume that $\phi \to \psi$ is a tautology, and let a be a truth assignment that satisfies ϕ. If a does not satisfy ψ, then we would have $\phi \to \psi \equiv T \to F \equiv F$. So, a must satisfy ψ. Since a was an arbitrary truth assignment that satisfies ϕ, $\phi \vdash \psi$. □

LEVEL 3

> 5. Determine if each of the following statements is a tautology, a contradiction, or neither. (i) $p \wedge p$; (ii) $p \wedge \neg p$; (iii) $(p \vee \neg p) \to (p \wedge \neg p)$; (iv) $\neg(p \vee q) \leftrightarrow (\neg p \wedge \neg q)$; (v) $p \to (\neg q \wedge r)$; (vi) $(p \leftrightarrow q) \to (p \to q)$

Solutions:

(i) If $p \equiv T$, then $p \wedge p \equiv T \wedge T \equiv T$. If $p \equiv F$, then $p \wedge p \equiv F \wedge F \equiv F$. **Neither**

(ii) $p \wedge \neg p \equiv F$. **Contradiction**

(iii) $(p \vee \neg p) \to (p \wedge \neg p) \equiv T \to F \equiv F$. **Contradiction**

(iv) Since $\neg(p \vee q) \equiv \neg p \wedge \neg q$ (De Morgan's law), $\neg(p \vee q) \leftrightarrow (\neg p \wedge \neg q)$ is a **Tautology**.

(v) If $p \equiv F$, then we have $p \to (\neg q \wedge r) \equiv F \to (\neg q \wedge r) \equiv T$. If $p \equiv T$ and $r \equiv F$, then we have $p \to (\neg q \wedge r) \equiv T \to (\neg q \wedge F) \equiv T \to F \equiv F$. **Neither**

(vi) Since $(p \leftrightarrow q) \equiv (p \to q) \wedge (q \to p)$ (by the law of the biconditional), we have that $(p \leftrightarrow q) \leftrightarrow [(p \to q) \wedge (q \to p)]$ is a tautology. In particular, $(p \leftrightarrow q) \vdash [(p \to q) \wedge (q \to p)]$. Since we also have $(p \to q) \wedge (q \to p) \vdash p \to q$, by transitivity of $\vdash$ (Problem 3 above), $(p \leftrightarrow q) \vdash p \to q$. Therefore, by Problem 4 above, $(p \leftrightarrow q) \to (p \to q)$ is a **Tautology**.

> 6. Verify all the rules of inference given in List 9.2.

Modus Ponens	Modus Tollens	Disjunctive Syllogism	Hypothetical Syllogism
$p \to q$	$p \to q$	$p \vee q$	$p \to q$
p	$\neg q$	$\neg p$	$q \to r$
q	$\neg p$	q	$p \to r$

Modus Ponens: This was done in Example 9.8.

Modus Tollens: Suppose that $p \to q \equiv T$ and $\neg q \equiv T$. Then $q \equiv F$, and therefore, we must have $p \equiv F$. So, $\neg p \equiv T$.

Disjunctive Syllogism: Suppose that $p \vee q \equiv T$ and $\neg p \equiv T$. Then $p \equiv F$, and therefore, we must have $q \equiv T$.

Hypothetical Syllogism: Suppose that $p \rightarrow q \equiv T$ and $q \rightarrow r \equiv T$. If $p \equiv T$, then since $p \rightarrow q \equiv T$, we must have $q \equiv T$. Since $q \rightarrow r \equiv T$, we must have $r \equiv T$. Since $p \equiv T$ and $r \equiv T$, we have $p \rightarrow r \equiv T$. If $p \equiv F$, then $p \rightarrow r \equiv T$.

Conjunctive Introduction	Disjunctive Introduction	Biconditional Introduction	Constructive Dilemma
p	p	$p \rightarrow q$	$p \rightarrow q$
q	---	$q \rightarrow p$	$r \rightarrow s$
---	$p \vee q$	---	$p \vee r$
$p \wedge q$		$p \leftrightarrow q$	$q \vee s$

Conjunctive Introduction: Suppose that $p \equiv T$ and $q \equiv T$. Then $p \wedge q \equiv T$.

Disjunctive Introduction: Suppose that $p \equiv T$. Then $p \vee q \equiv T$.

Biconditional Introduction: Suppose that $p \rightarrow q \equiv T$ and $q \rightarrow p \equiv T$. If $p \equiv T$, then since $p \rightarrow q \equiv T$, we must have $q \equiv T$. Since $p \equiv T$ and $q \equiv T$, we have $p \leftrightarrow q \equiv T$. If $p \equiv F$, then since $q \rightarrow p \equiv T$, we must have $q \equiv F$. Since $p \equiv F$ and $q \equiv F$, we have $p \leftrightarrow q \equiv T$.

Constructive Dilemma: Suppose that $p \rightarrow q \equiv T$, $r \rightarrow s \equiv T$, and $p \vee r \equiv T$. If $q \equiv F$ and $s \equiv F$, then since $p \rightarrow q \equiv T$, we must have $p \equiv F$. Since $r \rightarrow s \equiv T$, we must have $r \equiv F$. But then $p \vee r \equiv F$. So, there is no truth assignment satisfying $p \rightarrow q$, $r \rightarrow s$, and $p \vee r$ that will make $q \vee s \equiv F$.

Conjunctive Elimination	Disjunctive Resolution	Biconditional Elimination	Destructive Dilemma
$p \wedge q$	$p \vee q$	$p \leftrightarrow q$	$p \rightarrow q$
---	$\neg p \vee r$	---	$r \rightarrow s$
p	---	$p \rightarrow q$	$\neg q \vee \neg s$
	$q \vee r$		---
			$\neg p \vee \neg r$

Conjunctive Elimination: Suppose that $p \wedge q \equiv T$. Then $p \equiv T$ and $q \equiv T$. In particular, $p \equiv T$.

Disjunctive Resolution: Suppose that $p \vee q \equiv T$ and $\neg p \vee r \equiv T$. If $q \equiv F$ and $r \equiv F$, then since $p \vee q \equiv T$, we must have $p \equiv T$. Since $\neg p \vee r \equiv T$, we must have $\neg p \equiv T$. But then $p \wedge \neg p \equiv T$, which is impossible. So, there is no truth assignment satisfying $p \vee q$ and $\neg p \vee r$ that will make $q \vee r \equiv F$.

Biconditional Elimination: Suppose that $p \leftrightarrow q \equiv T$. If $p \rightarrow q \equiv F$, then $p \equiv T$ and $q \equiv F$. But then $p \leftrightarrow q \equiv F$. So, there is no truth assignment satisfying $p \leftrightarrow q$ that will make $p \rightarrow q \equiv F$.

Destructive Dilemma: Suppose that $p \rightarrow q \equiv T$, $r \rightarrow s \equiv T$, and $\neg q \vee \neg s \equiv T$. If $\neg p \equiv F$ and $\neg r \equiv F$, then $p \equiv T$ and $r \equiv T$. Since $p \rightarrow q \equiv T$, we must have $q \equiv T$. Since $r \rightarrow s \equiv T$, we must have $s \equiv T$. Then $\neg q \equiv F$ and $\neg s \equiv F$. Thus, $\neg q \vee \neg s \equiv F$. So, there is no truth assignment satisfying $p \rightarrow q$, $r \rightarrow s$, and $\neg q \vee \neg s$ that will make $\neg p \vee \neg r \equiv F$.

Level 4

7. Determine whether each of the following logical arguments is valid or invalid. If the argument is valid, provide a deduction. If the argument is invalid, provide a counterexample.

I	II	III	IV
$p \vee q$	$\neg(p \wedge q)$	$\neg p$	$p \to q$
q	q	$p \vee r$	$r \to \neg q$
$\overline{p}$	$\overline{\neg p}$	$q \to \neg r$	$\overline{p \to r}$
		$\overline{\neg q}$	

Solutions:

I. If we let $p \equiv F$ and $q \equiv T$, then $p \vee q \equiv F \vee T \equiv T$. So, we have found a truth assignment that makes the premises true and the conclusion false. Therefore, the argument is **invalid**.

II. Here is a derivation.

1	$\neg(p \wedge q)$	Premise
2	q	Premise
3	$\neg p \vee \neg q$	De Morgan's law (1)
4	$\neg(\neg q)$	Law of double negation (2)
5	$\neg p$	Disjunctive syllogism (3, 4)

Therefore, the argument is **valid**.

III. Here is a derivation.

1	$\neg p$	Premise
2	$p \vee r$	Premise
3	$q \to \neg r$	Premise
4	r	Disjunctive syllogism (2, 1)
5	$\neg(\neg r)$	Law of double negation (4)
6	$\neg q$	Modus tollens (3, 5)

Therefore, the argument is **valid**.

IV. If we let $p \equiv T$, $q \equiv T$ and $r \equiv F$, then $p \to q \equiv T \to T \equiv T$, $r \to \neg q \equiv F \to F \equiv T$, and $p \to r \equiv T \to F \equiv F$. So, we have found a truth assignment that makes the premises true and the conclusion false. Therefore, the argument is **invalid**.

8. Simplify each statement. (i) $p \vee (p \wedge \neg p)$; (ii) $(p \wedge q) \vee \neg p$; (iii) $\neg p \to (\neg q \to p)$; (iv) $(p \wedge \neg q) \vee p$; (v) $[(q \wedge p) \vee q] \wedge [(q \vee p) \wedge p]$

Solutions:

(i) $p \vee (p \wedge \neg p) \equiv p \vee F \equiv \boldsymbol{p}$.

(ii) $(p \wedge q) \vee \neg p \equiv (p \vee \neg p) \wedge (q \vee \neg p) \equiv T \wedge (q \vee \neg p) \equiv q \vee \neg p \equiv \boldsymbol{\neg p \vee q}$.

(iii) $\neg p \to (\neg q \to p) \equiv p \vee (\neg q \to p) \equiv p \vee (q \vee p) \equiv p \vee (p \vee q) \equiv (p \vee p) \vee q \equiv \boldsymbol{p \vee q}$.

(iv) $(p \wedge \neg q) \vee p \equiv \boldsymbol{p}$ (Absorption).

(v) $[(q \wedge p) \vee q] \wedge [(q \vee p) \wedge p] \equiv [(q \wedge p) \vee q] \wedge [(p \vee q) \wedge p] \equiv q \wedge p$ (Absorption) $\equiv \boldsymbol{p \wedge q}$.

LEVEL 5

9. Determine if the following logical argument is valid. If the argument is valid, provide a deduction. If the argument is invalid, provide a counterexample.

> If a piano has 88 keys, then the box is empty.
> If a piano does not have 88 keys, then paintings are white.
> If we are in immediate danger, then the box is not empty.
> Therefore, paintings are white or we are not in immediate danger.

Solution: Let p represent "A piano has 88 keys," let b represent "The box is empty," let w represent "Paintings are white," and let d represent "We are in immediate danger." We now give a deduction showing that the argument is valid.

1	$p \to b$	Premise
2	$\neg p \to w$	Premise
3	$d \to \neg b$	Premise
4	$\neg w \to \neg(\neg p)$	Law of the contrapositive (2)
5	$\neg w \to p$	Law of double negation (4)
6	$\neg w \to b$	Hypothetical syllogism (5, 1)
7	$\neg(\neg b) \to \neg d$	Law of the contrapositive (3)
8	$b \to \neg d$	Law of double negation (7)
9	$\neg w \to \neg d$	Hypothetical syllogism (6, 8)
10	$\neg(\neg w) \vee \neg d$	Law of the conditional (9)
11	$w \vee \neg d$	Law of double negation (10)

Therefore, the argument is **valid**.

10. Determine if the following logical argument is valid. If the argument is valid, provide a deduction. If the argument is invalid, provide a counterexample.

> Tangs have fangs or tings have wings.
> It is not the case that tangs have fangs and tings do not have wings.
> It is not the case that tangs do not have fangs and tings have wings.
> Therefore, tangs have fangs and either tings have wings or tangs do not have fangs.

Solution: Let f represent "Tangs have fangs," let w represent "Tings have wings." We now give a deduction showing that the argument is valid.

1	$t \lor w$	Premise
2	$\neg(t \land \neg w)$	Premise
3	$\neg(\neg t \land w)$	Premise
4	$\neg t \lor \neg(\neg w)$	De Morgan's law (2)
5	$\neg t \lor w$	Law of double negation (4)
6	$w \lor w$	Disjunctive resolution (1, 5)
7	w	Redundancy law (6)
8	$w \lor \neg t$	Disjunctive introduction (7)
9	$w \lor t$	Commutative law (1)
10	$\neg(\neg t) \lor \neg w$	De Morgan's law (3)
11	$t \lor \neg w$	Law of double negation (10)
12	$\neg w \lor t$	Commutative law (11)
13	$t \lor t$	Disjunctive resolution (9, 12)
14	t	Redundancy law (13)
15	$t \land (w \lor \neg t)$	Conjunctive introduction (14, 8)

Therefore, the argument is **valid**.

Problem Set 10

Level 1

1. For each set A below, evaluate (i) A^2; (ii) $\mathcal{P}(A)$; (iii) $^A A$

   1. $A = \emptyset$ 2. $A = \{\emptyset\}$ 3. $A = \{0,1\}$ 4. $A = \mathcal{P}(\{\emptyset\})$

Solutions:

(i) $\emptyset^2 = \emptyset \times \emptyset = \emptyset$.

   $\{\emptyset\}^2 = \{\emptyset\} \times \{\emptyset\} = \{(\emptyset, \emptyset)\}$.

   $\{0,1\}^2 = \{0,1\} \times \{0,1\} = \{(0,0), (0,1), (1,0), (1,1)\}$.

   Since $\mathcal{P}(\{\emptyset\}) = \{\emptyset, \{\emptyset\}\}$, we have

   $$\mathcal{P}(\{\emptyset\})^2 = \mathcal{P}(\{\emptyset\}) \times \mathcal{P}(\{\emptyset\}) = \{(\emptyset, \emptyset), (\emptyset, \{\emptyset\}), (\{\emptyset\}, \emptyset), (\{\emptyset\}, \{\emptyset\})\}.$$

(ii) $\mathcal{P}(\emptyset) = \{\emptyset\}$; $\mathcal{P}(\{\emptyset\}) = \{\emptyset, \{\emptyset\}\}$

   $\mathcal{P}(\{0,1\}) = \{\emptyset, \{0\}, \{1\}, \{0,1\}\}$

   Since $\mathcal{P}(\{\emptyset\}) = \{\emptyset, \{\emptyset\}\}$, we have $\mathcal{P}(\mathcal{P}(\{\emptyset\})) = \{\emptyset, \{\emptyset\}, \{\{\emptyset\}\}, \{\emptyset, \{\emptyset\}\}\}$.

(iii) $^\emptyset \emptyset = \emptyset$

   $^{\{\emptyset\}}\{\emptyset\} = \{\{(\emptyset, \emptyset)\}\}$

   $^{\{0,1\}}\{0,1\} = \{\{(0,0), (1,0)\}, \{(0,0), (1,1)\}, \{(0,1), (1,0)\}, \{(0,1), (1,1)\}\}$

   $^{\mathcal{P}(\{\emptyset\})}\mathcal{P}(\{\emptyset\}) = {}^{\{\emptyset, \{\emptyset\}\}}\{\emptyset, \{\emptyset\}\}$. So, we get

   $$\{\{(\emptyset, \emptyset), (\{\emptyset\}, \emptyset)\}, \{(\emptyset, \emptyset), (\{\emptyset\}, \{\emptyset\})\}, \{(\emptyset, \{\emptyset\}), (\{\emptyset\}, \emptyset)\}, \{(\emptyset, \{\emptyset\}), (\{\emptyset\}, \{\emptyset\})\}\}.$$

2. Find all partitions of the three element set $\{a, b, c\}$ and the four element set $\{a, b, c, d\}$.

Solution: The partitions of $\{a,b,c\}$ are $\{\{a\},\{b\},\{c\}\}$, $\{\{a\},\{b,c\}\}$, $\{\{b\},\{a,c\}\}$, $\{\{c\},\{a,b\}\}$, and $\{\{a,b,c\}\}$.

The partitions of $\{a,b,c,d\}$ are $\{\{a\},\{b\},\{c\},\{d\}\}$, $\{\{a\},\{b\},\{c,d\}\}$, $\{\{a\},\{c\},\{b,d\}\}$, $\{\{a\},\{d\},\{b,c\}\}$, $\{\{b\},\{c\},\{a,d\}\}$, $\{\{b\},\{d\},\{a,c\}\}$, $\{\{c\},\{d\},\{a,b\}\}$, $\{\{a,b\},\{c,d\}\}$, $\{\{a,c\},\{b,d\}\}$, $\{\{a,d\},\{b,c\}\}$, $\{\{a,b,c\},\{d\}\}$, $\{\{a,b,d\},\{c\}\}$, $\{\{a,c,d\},\{b\}\}$, $\{\{b,c,d\},\{a\}\}$, and $\{\{a,b,c,d\}\}$.

Level 2

3. For $a, b \in \mathbb{N}$, we will say that a divides b, written $a|b$, if there is a natural number k such that $b = ak$. Notice that $|$ is a relation on $\mathbb{N}$. Prove that $(\mathbb{N}, |)$ is a partially ordered set, but it is not a linearly ordered set.

Proof: If $a \in \mathbb{N}$ then $a = 1a$, so that $a|a$. Therefore, $|$ is reflexive.

If $a|b$ and $b|a$, then there are natural numbers j and k such that $b = ja$ and $a = kb$. If $a = 0$, then $b = j \cdot 0 = 0$, and so, $a = b$. Suppose $a \neq 0$. Then we have $a = k(ja) = (kj)a$. Therefore, $(kj - 1)a = (kj)a - 1a = 0$. So, $kj - 1 = 0$, and therefore, $kj = 1$. So, $k = j = 1$. Thus, $b = ja = 1a = a$. Therefore, $|$ is antisymmetric.

If $a|b$ and $b|c$, then there are natural numbers j and k such that $b = ja$ and $c = kb$. Then $c = kb = k(ja) = (kj)a$. Since $\mathbb{N}$ is closed under multiplication, $kj \in \mathbb{N}$. So, $a|c$. Therefore, $|$ is transitive.

Since $|$ is reflexive, antisymmetric, and transitive on $\mathbb{N}$, $(\mathbb{N}, |)$ is a partially ordered set.

Since 2 and 3 do not divide each other, $(\mathbb{N}, |)$ is **not** linearly ordered. □

4. Prove that for each $n \in \mathbb{Z}^+$, $\equiv_n$ (see part 3 of Example 10.8) is an equivalence relation on $\mathbb{Z}$.

Proof: Let $a \in \mathbb{Z}$. Then $a - a = 0 = n \cdot 0$. So, $n|a - a$. Therefore, $a \equiv_n a$, and so, $\equiv_n$ is reflexive.

Let $a, b \in \mathbb{Z}$ and suppose that $a \equiv_n b$. Then $n|b - a$. So, there is $k \in \mathbb{Z}$ such that $b - a = nk$. Thus, $a - b = -(b - a) = -nk = n(-k)$. Since $k \in \mathbb{Z}$, $-k \in \mathbb{Z}$. So, $n|a - b$, and therefore, $b \equiv_n a$. So, $\equiv_n$ is symmetric.

Let $a, b, c \in \mathbb{Z}$ with $a \equiv_n b$ and $b \equiv_n c$. Then $n|b - a$ and $n|c - b$. So, there are $j, k \in \mathbb{Z}$ such that $b - a = nj$ and $c - b = nk$. So, $c - a = (c - b) + (b - a) = nk + nj = n(k + j)$. Since $\mathbb{Z}$ is closed under addition, $k + j \in \mathbb{Z}$. Therefore, $n|c - a$. So, $a \equiv_n c$. Thus, $\equiv_n$ is transitive.

Since $\equiv_n$ is reflexive, symmetric, and transitive, $\equiv_n$ is an equivalence relation on $\mathbb{Z}$. □

5. Let A, B, and C be sets. Prove the following: (i) If $A \subseteq B$, then $A \preccurlyeq B$. (ii) $\preccurlyeq$ is transitive. (iii) $\prec$ is transitive. (iv) If $A \preccurlyeq B$ and $B \prec C$, then $A \prec C$. (v) If $A \prec B$ and $B \preccurlyeq C$, then $A \prec C$.

Proofs:

(i) Let $A \subseteq B$ and define $f: A \to B$ by $f(x) = x$. Then f is clearly an injection, and so, $A \preccurlyeq B$. □

(ii) Suppose that $A \preccurlyeq B$ and $B \preccurlyeq C$. Then there are functions $f: A \hookrightarrow B$ and $g: B \hookrightarrow C$. By Theorem 10.3, $g \circ f: A \hookrightarrow C$. So, $A \preccurlyeq C$. Therefore, $\preccurlyeq$ is transitive. □

(iii) Suppose that $A \prec B$ and $B \prec C$. Then $A \preccurlyeq B$ and $B \preccurlyeq C$. By (ii), $A \preccurlyeq C$. Assume toward contradiction that $A \sim C$. Since $\sim$ is symmetric, $C \sim A$. In particular, $C \preccurlyeq A$. Since $C \preccurlyeq A$ and $A \preccurlyeq B$, by (ii), $C \preccurlyeq B$. Since $B \preccurlyeq C$ and $C \preccurlyeq B$, by the Cantor-Schroeder-Bernstein Theorem, $B \sim C$, contradicting $B \prec C$. It follows that $A \not\sim C$, and thus, $A \prec C$. □

(iv) Suppose that $A \preccurlyeq B$ and $B \prec C$. Then $B \preccurlyeq C$. By (ii), $A \preccurlyeq C$. Assume toward contradiction that $A \sim C$. The rest of the argument is the same as (iii). □

(v) Suppose that $A \prec B$ and $B \preccurlyeq C$. Then $A \preccurlyeq B$. By (ii), $A \preccurlyeq C$. Assume toward contradiction that $A \sim C$. Since $\sim$ is symmetric, $C \sim A$. In particular, $C \preccurlyeq A$. Since $B \preccurlyeq C$ and $C \preccurlyeq A$, by (ii), $B \preccurlyeq A$. Since $A \preccurlyeq B$ and $B \preccurlyeq A$, by the Cantor-Schroeder-Bernstein Theorem, $A \sim B$, contradicting $A \prec B$. It follows that $A \nsim C$, and thus, $A \prec C$. □

6. Let A and B be sets such that $A \subseteq B$. Prove that $\mathcal{P}(A) \preccurlyeq \mathcal{P}(B)$.

Proof: Suppose that $A \subseteq B$. We show that $\mathcal{P}(A) \subseteq \mathcal{P}(B)$. Let $X \in \mathcal{P}(A)$. Then $X \subseteq A$. Since $X \subseteq A$ and $A \subseteq B$, and $\subseteq$ is transitive (Theorem 2.3), we have $X \subseteq B$. Therefore, $X \in \mathcal{P}(B)$. Since X was an arbitrary element of $\mathcal{P}(A)$, we have shown that every element of $\mathcal{P}(A)$ is an element of $\mathcal{P}(B)$. Therefore, $\mathcal{P}(A) \subseteq \mathcal{P}(B)$. By Problem 5 (part (i)), $\mathcal{P}(A) \preccurlyeq \mathcal{P}(B)$. □

LEVEL 3

7. For $f, g \in {}^{\mathbb{R}}\mathbb{R}$, define $f \preccurlyeq g$ if and only if for all $x \in \mathbb{R}$, $f(x) \leq g(x)$. Is $({}^{\mathbb{R}}\mathbb{R}, \preccurlyeq)$ a poset? Is it a linearly ordered set? What if we replace $\preccurlyeq$ by $\preccurlyeq^*$, where $f \preccurlyeq^* g$ if and only if there is an $x \in \mathbb{R}$ such that $f(x) \leq g(x)$?

Solution: If $f \in {}^{\mathbb{R}}\mathbb{R}$, then for all $x \in \mathbb{R}$, $f(x) = f(x)$. So, $f \preccurlyeq f$, and therefore, $\preccurlyeq$ is reflexive.

Let $f, g \in {}^{\mathbb{R}}\mathbb{R}$ with $f \preccurlyeq g$ and $g \preccurlyeq f$. Then for all $x \in \mathbb{R}$, $f(x) \leq g(x)$ and $g(x) \leq f(x)$. So, $f = g$, and therefore, $\preccurlyeq$ is antisymmetric.

Let $f, g, h \in {}^{\mathbb{R}}\mathbb{R}$ with $f \preccurlyeq g$ and $g \preccurlyeq h$. Then for all $x \in \mathbb{R}$, $f(x) \leq g(x)$ and $g(x) \leq h(x)$. So, by the transitivity of $\leq$, for all $x \in \mathbb{R}$, $f(x) \leq h(x)$. Thus, $f \preccurlyeq h$, and therefore, $\preccurlyeq$ is transitive.

Since $\preccurlyeq$ is reflexive, antisymmetric, and transitive, $({}^{\mathbb{R}}\mathbb{R}, \preccurlyeq)$ is a poset.

Let $f(x) = x$ and $g(x) = x^2$. Then $f(2) = 2$ and $g(2) = 4$. So, $f(2) < g(2)$. Therefore, $g \npreccurlyeq f$. We also have $f\left(\frac{1}{2}\right) = \frac{1}{2}$ and $g\left(\frac{1}{2}\right) = \frac{1}{4}$. So, $g\left(\frac{1}{2}\right) < f\left(\frac{1}{2}\right)$. Therefore, $f \npreccurlyeq g$. So, f and g are incomparable with respect to $\preccurlyeq$. Therefore, $({}^{\mathbb{R}}\mathbb{R}, \preccurlyeq)$ is **not** a linearly ordered set.

The same example from the last paragraph gives us $f \preccurlyeq^* g$ and $g \preccurlyeq^* f$. But $f \neq g$. So, $\preccurlyeq^*$ is **not** antisymmetric, and therefore, $({}^{\mathbb{R}}\mathbb{R}, \preccurlyeq^*)$ is **not** a poset.

8. Prove that the function $f: \mathbb{N} \to \mathbb{Z}$ defined by $f(n) = \begin{cases} \frac{n}{2} & \text{if } n \text{ is even} \\ -\frac{n+1}{2} & \text{if } n \text{ is odd} \end{cases}$ is a bijection.

Proof: First note that if n is even, then there is $k \in \mathbb{Z}$ with $n = 2k$, and so, $\frac{n}{2} = \frac{2k}{2} = k \in \mathbb{Z}$, and if n is odd, there is $k \in \mathbb{Z}$ with $n = 2k + 1$, and so, $-\frac{n+1}{2} = -\frac{(2k+1)+1}{2} = -\frac{2k+2}{2} = -\frac{2(k+1)}{2} = -(k+1) \in \mathbb{Z}$. So, f does take each natural number to an integer.

Now, suppose that $n, m \in \mathbb{N}$ with $f(n) = f(m)$. If n and m are both even, we have $\frac{n}{2} = \frac{m}{2}$, and so, $2 \cdot \frac{n}{2} = 2 \cdot \frac{m}{2}$. Thus, $n = m$. If n and m are both odd, we have $-\frac{n+1}{2} = -\frac{m+1}{2}$, and so, $\frac{n+1}{2} = \frac{m+1}{2}$. Thus, $2 \cdot \frac{n+1}{2} = 2 \cdot \frac{m+1}{2}$. So, $n+1 = m+1$, and therefore, $n = m$. If n is even and m is odd, then we have $\frac{n}{2} = -\frac{m+1}{2}$. So, $2 \cdot \frac{n}{2} = 2\left(-\frac{m+1}{2}\right)$. Therefore, $n = -(m+1)$. Since $m \in \mathbb{N}$, $m \geq 0$. So, $m + 1 \geq 1$. Therefore, $n = -(m+1) \leq -1$, contradicting $n \in \mathbb{N}$. So, it is impossible for n to be even, m to be odd, and $f(n) = f(m)$. Similarly, we cannot have n odd and m even. So, f is an injection.

Now, let $k \in \mathbb{Z}$. If $k \geq 0$, then $2k \in \mathbb{N}$ and $f(2k) = \frac{2k}{2} = k$. If $k < 0$, then $-2k > 0$, and so, we have $-2k - 1 \in \mathbb{N}$. Then $f(-2k - 1) = -\frac{(-2k-1)+1}{2} = -\frac{-2k}{2} = k$. So, f is a surjection.

Since f is both an injection and a surjection, f is a bijection. □

9. Define $\mathcal{P}_k(\mathbb{N})$ for each $k \in \mathbb{N}$ by $\mathcal{P}_0(\mathbb{N}) = \mathbb{N}$ and $\mathcal{P}_{k+1}(\mathbb{N}) = \mathcal{P}(\mathcal{P}_k(\mathbb{N}))$ for $k > 0$. Find a set B such that for all $k \in \mathbb{N}$, $\mathcal{P}_k(\mathbb{N}) \prec B$.

Solution: Let $B = \cup\{\mathcal{P}_n(\mathbb{N}) \mid n \in \mathbb{N}\}$. Let $k \in \mathbb{N}$. Since $\mathcal{P}_k(\mathbb{N}) \subseteq B$, by Problem 5 (part (i)), $\mathcal{P}_k(\mathbb{N}) \preccurlyeq B$. Since k was arbitrary, we have $\mathcal{P}_k(\mathbb{N}) \preccurlyeq B$ for all $k \in \mathbb{N}$. Again, let $k \in \mathbb{N}$. We have $\mathcal{P}_k(\mathbb{N}) \prec \mathcal{P}_{k+1}(\mathbb{N})$ and $\mathcal{P}_{k+1}(\mathbb{N}) \preccurlyeq B$. By Problem 5 (part (v)), $\mathcal{P}_k(\mathbb{N}) \prec B$. Since $k \in \mathbb{N}$ was arbitrary, we have shown that for all $k \in \mathbb{N}$, $\mathcal{P}_k(\mathbb{N}) \prec B$.

10. Prove that if $A \sim B$ and $C \sim D$, then $A \times C \sim B \times D$.

Proof: Suppose that $A \sim B$ and $C \sim D$. Then there exist bijections $h: A \to B$ and $k: C \to D$. Define $f: A \times C \to B \times D$ by $f(a, c) = (h(a), k(c))$.

Suppose $(a, c), (a', c') \in A \times C$ with $f((a, c)) = f((a', c'))$. Then $(h(a), k(c)) = (h(a'), k(c'))$. So, $h(a) = h(a')$ and $k(c) = k(c')$. Since h is an injection, $a = a'$. Since k is an injection, $c = c'$. Since $a = a'$ and $c = c'$, $(a, c) = (a', c')$. Since $(a, c), (a', c') \in A \times C$ were arbitrary, f is an injection.

Now, let $(b, d) \in B \times D$. Since h and k are bijections, h^{-1} and k^{-1} exist. Let $a = h^{-1}(b)$, $c = k^{-1}(d)$. Then $f(a, c) = (h(a), k(c)) = (h(h^{-1}(b)), k(k^{-1}(d))) = (b, d)$. Since $(b, d) \in B \times D$ was arbitrary, f is a surjection.

Since f is both an injection and a surjection, $A \times C \sim B \times D$. □

LEVEL 4

11. Define a partition P of $\mathbb{N}$ such that $P \sim \mathbb{N}$ and for each $X \in P$, $X \sim \mathbb{N}$.

Proof: For each $n \in \mathbb{N}$, let P_n be the set of natural numbers ending with exactly n zeros and let $\boldsymbol{P} = \{P_n \mid n \in \mathbb{N}\}$. For example, $5231 \in P_0$, $0 \in P_1$, and $26{,}200 \in P_2$. Let's define $\widetilde{m,n}$ to be the natural number consisting of m 1's followed by n 0's. For example, $\widetilde{3,0} = 111$ and $\widetilde{2,5} = 1{,}100{,}000$. For each $n \in \mathbb{N}$, $\{\widetilde{m,n} \mid m \in \mathbb{N}\} \subseteq P_n$ showing that each P_n is equinumerous to $\mathbb{N}$. Also, if $k \in P_n \cap P_m$, then k ends with exactly n zeros and exactly m zeros, and so, $n = m$. Therefore, $\boldsymbol{P}$ is pairwise disjoint. This also shows that the function $f: \mathbb{N} \to \boldsymbol{P}$ defined by $f(n) = P_n$ is a bijection. So, $\boldsymbol{P} \sim \mathbb{N}$. Finally, if $k \in \mathbb{N}$, then there is $n \in \mathbb{N}$ such that k ends with exactly n zeros. So, $\bigcup \boldsymbol{P} = \mathbb{N}$. $\square$

12. Prove that a countable union of countable sets is countable.

Proof: For each $n \in \mathbb{N}$, let A_n be a countable set. By replacing each A_n by $A_n \times \{n\}$, we can assume that $\{A_n \mid n \in \mathbb{N}\}$ is a pairwise disjoint collection of sets ($A_n \sim A_n \times \{n\}$ via the bijection f sending x to (x, n)). By Problem 11, there is a partition $\boldsymbol{P}$ of $\mathbb{N}$ such that $\boldsymbol{P} \sim \mathbb{N}$ and for each $X \in \boldsymbol{P}$, $X \sim \mathbb{N}$. Let's say $\boldsymbol{P} = \{P_n \mid n \in \mathbb{N}\}$. Since each A_n is countable, for each $n \in \mathbb{N}$ there are injective functions $f_n: A_n \to P_n$. Define $f: \bigcup\{A_n \mid n \in \mathbb{N}\} \to \mathbb{N}$ by $f(x) = f_n(x)$ if $x \in A_n$.

Since $\{A_n \mid n \in \mathbb{N}\}$ is pairwise disjoint, f is well-defined.

Suppose that $x, y \in \bigcup\{A_n \mid n \in \mathbb{N}\}$ with $f(x) = f(y)$. There exist $n, m \in \mathbb{N}$ such that $x \in A_n$ and $y \in A_m$. So, $f(x) = f_n(x) \in P_n$ and $f(y) = f_m(y) \in P_m$. Since $f(x) = f(y)$, we have $f_n(x) = f_m(y)$. Since for $n \neq m$, $P_n \cap P_m = \emptyset$, we must have $n = m$. So, we have $f_n(x) = f_n(y)$. Since f_n is injective, $x = y$. Since $x, y \in \bigcup\{A_n \mid n \in \mathbb{N}\}$ were arbitrary, f is an injective function. Therefore, $\bigcup\{A_n \mid n \in \mathbb{N}\}$ is countable. $\square$

13. Let A and B be sets such that $A \sim B$. Prove that $\mathcal{P}(A) \sim \mathcal{P}(B)$.

Proof: Suppose that $A \sim B$. Then there exists a bijection $h: A \to B$. Define $F: \mathcal{P}(A) \to \mathcal{P}(B)$ by $F(X) = \{h(a) \mid a \in X\}$ for each $X \in \mathcal{P}(A)$.

Suppose $X, Y \in \mathcal{P}(A)$ with $F(X) = F(Y)$. Let $a \in X$. Then $h(a) \in F(X)$. Since $F(X) = F(Y)$, $h(a) \in F(Y)$. So, there is $b \in Y$ such that $h(a) = h(b)$. Since h is injective, $a = b$. So, $a \in Y$. Since $a \in X$ was arbitrary, $X \subseteq Y$. By a symmetrical argument, $Y \subseteq X$. Therefore, $X = Y$. Since $X, Y \in \mathcal{P}(A)$ were arbitrary, F is injective.

Let $Y \in \mathcal{P}(B)$ and let $X = \{a \in A \mid h(a) \in Y\}$. Then $b \in F(X)$ if and only if $b = h(a)$ for some $a \in X$ if and only if $b \in Y$ (because h is surjective). So, $F(X) = Y$. Since $Y \in \mathcal{P}(B)$ was arbitrary, F is surjective.

Since F is injective and surjective, $\mathcal{P}(A) \sim \mathcal{P}(B)$. $\square$

14. Prove the following: (i) $\mathbb{N} \times \mathbb{N} \sim \mathbb{N}$. (ii) $\mathbb{Q} \sim \mathbb{N}$. (iii) Any two intervals of real numbers are equinumerous (including $\mathbb{R}$ itself). (iv) $^\mathbb{N}\mathbb{N} \sim \mathcal{P}(\mathbb{N})$.

Proofs:

(i) $\mathbb{N} \times \mathbb{N} = \bigcup\{\mathbb{N} \times \{n\} \mid n \in \mathbb{N}\}$. This is a countable union of countable sets. By Problem 12, $\mathbb{N} \times \mathbb{N}$ is countable. $\square$

(ii) $\mathbb{Q}^+ = \left\{\frac{a}{b} \mid a \in \mathbb{N} \wedge b \in \mathbb{N}^+\right\} = \bigcup\left\{\left\{\frac{a}{b} \mid a \in \mathbb{N}\right\} \mid b \in \mathbb{N}^+\right\}$. This is a countable union of countable sets. By Problem 12, $\mathbb{Q}^+$ is countable. Now, $\mathbb{Q} = \mathbb{Q}^+ \cup \{0\} \cup \mathbb{Q}^-$, where $\mathbb{Q}^- = \{q \in \mathbb{Q} \mid -q \in \mathbb{Q}^+\}$. This is again a countable union of countable sets, thus countable. So, $\mathbb{Q} \sim \mathbb{N}$. □

(iii) The function $f: \mathbb{R} \to (0, \infty)$ defined by $f(x) = 2^x$ is a bijection. So, $\mathbb{R} \sim (0, \infty)$. The function $g: (0, \infty) \to (0, 1)$ defined by $g(x) = \frac{1}{x^2+1}$ is a bijection. So, $(0, \infty) \sim (0, 1)$. If $a, b \in \mathbb{R}$, the function $h: (0, 1) \to (a, b)$ defined by $h(x) = (b-a)x + a$ is a bijection. So, $(0, 1) \sim (a, b)$. It follows that all bounded open intervals are equinumerous with each other and $\mathbb{R}$.

We have, $[a, b] \subseteq (a - 1, b + 1) \sim (a, b) \subseteq [a, b] \subseteq [a, b]$ and $(a, b) \subseteq (a, b] \subseteq [a, b]$. It follows that all bounded intervals are equinumerous with each other and $\mathbb{R}$.

We also have the following.
$$(a, \infty) \subseteq [a, \infty) \subseteq \mathbb{R} \sim (a, a+1) \subseteq (a, \infty)$$
$$(-\infty, b) \subseteq (-\infty, b] \subseteq \mathbb{R} \sim (b-1, b) \subseteq (-\infty, b)$$

Therefore, all unbounded intervals are equinumerous with $\mathbb{R}$. It follows that any two intervals of real numbers are equinumerous. □

(iv) $^{\mathbb{N}}\mathbb{N} \subseteq \mathcal{P}(\mathbb{N} \times \mathbb{N})$ by the definition of $^{\mathbb{N}}\mathbb{N}$. So, $^{\mathbb{N}}\mathbb{N} \preccurlyeq \mathcal{P}(\mathbb{N} \times \mathbb{N})$ by Problem 5 (part (i)). By (i) above, $\mathbb{N} \times \mathbb{N} \sim \mathbb{N}$. So, by Problem 13, $\mathcal{P}(\mathbb{N} \times \mathbb{N}) \sim \mathcal{P}(\mathbb{N})$. Therefore, $\mathcal{P}(\mathbb{N} \times \mathbb{N}) \preccurlyeq \mathcal{P}(\mathbb{N})$. Since $\preccurlyeq$ is transitive, $^{\mathbb{N}}\mathbb{N} \preccurlyeq \mathcal{P}(\mathbb{N})$.

Now, $\mathcal{P}(\mathbb{N}) \sim {}^{\mathbb{N}}\{0, 1\}$ (see Example 10.18 (part 5)). So, $\mathcal{P}(\mathbb{N}) \preccurlyeq {}^{\mathbb{N}}\{0, 1\}$. Also, $^{\mathbb{N}}\{0, 1\} \subseteq {}^{\mathbb{N}}\mathbb{N}$, and so, by Problem 5 (part (i)), $^{\mathbb{N}}\{0, 1\} \preccurlyeq {}^{\mathbb{N}}\mathbb{N}$. Since $\preccurlyeq$ is transitive, $\mathcal{P}(\mathbb{N}) \preccurlyeq {}^{\mathbb{N}}\mathbb{N}$.

By the Cantor-Schroeder-Bernstein Theorem, $^{\mathbb{N}}\mathbb{N} \sim \mathcal{P}(\mathbb{N})$. □

Notes: (1) In the proof of (iii), we used the fact that equinumerosity is an equivalence relation, the Cantor-Schroeder-Bernstein Theorem, and Problem 5 many times without mention. For example, we have $\mathbb{R} \sim (0, \infty)$ and $(0, \infty) \sim (0, 1)$. So, by the transitivity of $\sim$, we have $\mathbb{R} \sim (0, 1)$. As another example, the sequence $(a, \infty) \subseteq [a, \infty) \subseteq \mathbb{R} \sim (a, a+1) \subseteq (a, \infty)$ together with Problem 5 gives us that $(a, \infty) \preccurlyeq \mathbb{R}$ and $\mathbb{R} \preccurlyeq (a, \infty)$. By the Cantor-Schroeder-Bernstein Theorem, $(a, \infty) \sim \mathbb{R}$.

(2) Once we showed that for all $a, b \in \mathbb{R}$, $(0, 1) \sim (a, b)$, it follows from the fact that $\sim$ is an equivalence relation that any two bounded open intervals are equinumerous. Indeed, if (a, b) and (c, d) are bounded open intervals, then $(0, 1) \sim (a, b)$ and $(0, 1) \sim (c, d)$. By the symmetry of $\sim$, we have $(a, b) \sim (0, 1)$, and finally, by the transitivity of $\sim$, we have $(a, b) \sim (c, d)$.

(3) It's easy to prove that two specific intervals of real numbers are equinumerous using just the fact that any two bounded open intervals are equinumerous with each other, together with the fact that $\mathbb{R} \sim (0, 1)$. For example, to show that $[3, \infty)$ is equinumerous with $(-2, 5]$, simply consider the following sequence: $[3, \infty) \subseteq \mathbb{R} \sim (0, 1) \sim (-2, 5) \subseteq (-2, 5] \subseteq (-2, 6) \sim (3, 4) \subseteq [3, \infty)$.

15. Prove that $\{A \in \mathcal{P}(\mathbb{N}) \mid A \text{ is infinite}\}$ is uncountable.

Proof: We first show that $X = \{A \in \mathcal{P}(\mathbb{N}) \mid A \text{ is finite}\}$ is countable. For each $n \in \mathbb{N}$, let $A_n = \{A \in \mathcal{P}(\mathbb{N}) \mid |A| \leq n\}$. Since $X = \bigcup\{A_n \mid n \in \mathbb{N}\}$, by Problem 12, it suffices to show that for each $n \in \mathbb{N}$, A_n is countable. We show this by induction on $n \in \mathbb{N}$. $A_0 = \{\emptyset\}$, which is certainly countable. $\{\{n\} \mid n \in \mathbb{N}\}$ is clearly equinumerous to $\mathbb{N}$ via the function sending $\{n\}$ to n. Therefore, we see that $A_1 = A_0 \cup \{\{n\} \mid n \in \mathbb{N}\}$ is countable. Let $k \in \mathbb{N}$ and assume that A_k is countable. For each $n \in \mathbb{N}$, the set $B_k^n = \{A \cup \{n\} \mid A \in A_k\}$ is countable. By Problem 12, the set $B_{k+1} = \bigcup\{B_k^n \mid n \in \mathbb{N}\}$ is countable. So, $A_{k+1} = A_0 \cup B_{k+1}$ is countable. By the principle of mathematical induction, for each $n \in \mathbb{N}$, A_n is countable. It follows that $X = \{A \in \mathcal{P}(\mathbb{N}) \mid A \text{ is finite}\}$ is countable.

Let $Y = \{A \in \mathcal{P}(\mathbb{N}) \mid A \text{ is infinite}\}$. Since every subset of $\mathbb{N}$ is either finite or infinite, $\mathcal{P}(\mathbb{N}) = X \cup Y$. If Y were countable, then since X is countable, by Problem 12, $\mathcal{P}(\mathbb{N})$ would be countable, which we know it is not. Therefore, Y is uncountable. □

Note: Computing A_1 in the proof above was not necessary. $B_0^n = \{A \cup \{n\} \mid A \in A_0\} = \{\{n\}\}$. Therefore, $B_1 = \bigcup\{B_0^n \mid n \in \mathbb{N}\} = \{\{n\} \mid n \in \mathbb{N}\}$. So, $A_1 = A_0 \cup B_1 = A_0 \cup \{\{n\} \mid n \in \mathbb{N}\}$. This is the same set that we wrote out explicitly in the proof.

16. For $f, g \in {}^\mathbb{N}\mathbb{N}$, define $f <^* g$ if and only if there is $n \in \mathbb{N}$ such that for all $m > n$, $f(m) < g(m)$.
    (i) Is $({}^\mathbb{N}\mathbb{N}, <^*)$ a strict poset? (ii) Is $({}^\mathbb{N}\mathbb{N}, <^*)$ a strict linearly ordered set? (iii) Let $\mathcal{F} = \{f_n : \mathbb{N} \to \mathbb{N} \mid n \in \mathbb{N}\}$ be a countable set of functions. Must there be a function $g \in {}^\mathbb{N}\mathbb{N}$ such that for all $n \in \mathbb{N}$, $f_n <^* g$?

Solutions:

(i) If $f \in {}^\mathbb{N}\mathbb{N}$, then for all $n \in \mathbb{N}$, $f(n) = f(n)$. So, $f \not<^* f$, and therefore, $<^*$ is antireflexive.

Let $f, g \in {}^\mathbb{N}\mathbb{N}$ with $f <^* g$ and $g <^* f$. Since $f <^* g$, there is $n_1 \in \mathbb{N}$ such that for all $m > n_1$, $f(m) < g(m)$. Since $g <^* f$, there is $n_2 \in \mathbb{N}$ such that for all $m > n_2$, $g(m) < f(m)$. Let $n = \max\{n_1, n_2\}$. Then $f(n+1) < g(n+1)$ and $g(n+1) < f(n+1)$. By the transitivity of $<$, we have $f(n+1) < f(n+1)$, a contradiction. So, antisymmetry holds (vacuously).

Let $f, g, h \in {}^\mathbb{N}\mathbb{N}$ with $f <^* g$ and $g <^* h$. Since $f <^* g$, there is $n_1 \in \mathbb{N}$ such that for all $m > n_1$, $f(m) < g(m)$. Since $g <^* h$, there is $n_2 \in \mathbb{N}$ such that for all $m > n_2$, $g(m) < h(m)$. Let $n = \max\{n_1, n_2\}$. Then for $m > n$, we have $f(m) < g(m)$ and $g(m) < h(m)$. By the transitivity of $<$, for $m > n$, we have $f(m) < h(m)$. So, $f <^* h$. Therefore, $<^*$ is transitive.

Since $<^*$ is antireflexive, antisymmetric, and transitive, $({}^\mathbb{N}\mathbb{N}, <^*)$ is a strict poset.

(ii) Let $f(n) = \frac{1}{2}$ and $g(n) = \begin{cases} 0 & \text{if } n \text{ is even} \\ 1 & \text{if } n \text{ is odd} \end{cases}$. Then for each $k \in \mathbb{N}$, $g(2k) = 0 < \frac{1}{2} = f(2k)$. So, $f \not<^* g$. Also, for each $k \in \mathbb{N}$, $g(2k+1) = 1 > \frac{1}{2} = f(2k+1)$. Therefore, $g \not<^* f$. So, f and g are incomparible with respect to $<^*$. Therefore, $({}^\mathbb{N}\mathbb{N}, <^*)$ is **not** a strict linearly ordered set.

(iii) Let $\mathcal{F} = \{f_n: \mathbb{N} \to \mathbb{N} \mid n \in \mathbb{N}\}$ and define $g: \mathbb{N} \to \mathbb{N}$ by $g(k) = \max\{f_n(k) + 1 \mid n \leq k\}$.

Let $n \in \mathbb{N}$. If $m > n - 1$, then we have $f_n(m) < f_n(m) + 1 \leq g(m)$. So, $f_n <^* g$. It follows that for all $n \in \mathbb{N}$, $f_n <^* g$.

Notes: (1) To better understand the definition of $<^*$, let's look at an example. Define $f, g: \mathbb{N} \to \mathbb{N}$ by $f(n) = n + 100$ and $g(n) = 2^n$. Observe that $f(m) < g(m)$ for $m > 6$. It follows that $f <^* g$.

Note that $f(1) = 101$ and $g(1) = 2^1 = 2$, so that $f(1) > g(1)$. So, it's not true that $f(n) < g(n)$ for all n. The definition of $<^*$ allows for this. $f <^* g$ means that the values of f are *eventually* less than the values of g. The expression $f <^* g$ is usually read as "f is **dominated** by g" or "g **dominates** f."

(2) Consider the family $\mathcal{F} = \{f_n: \mathbb{N} \to \mathbb{N} \mid n \in \mathbb{N}\}$, where $f_n(k) = n$ for all $k \in \mathbb{N}$. For each $n \in \mathbb{N}$, f_n is a constant function. For example, f_0 is the function which gives an output of 0 for each natural number input. You can visualize this constant function as dots along the x-axis, as shown in the figure to the right. The figure also shows the functions f_1, f_2, and f_3. There is no function g such that for all $n \in \mathbb{N}$, $f_n(k) < g(k)$ for all $k \in \mathbb{N}$. However, there are functions which dominate every f_n. For example, let $g: \mathbb{N} \to \mathbb{N}$ be defined by $g(n) = n$. You can visualize g as the dots along the diagonal ray shown in the figure to the right. We see that $f_0(k) < g(k)$ for all $k > 0$. We also see that $f_1(k) < g(k)$ for all $k > 1$. In general, $f_n(k) < g(k)$ for all $k > n$.

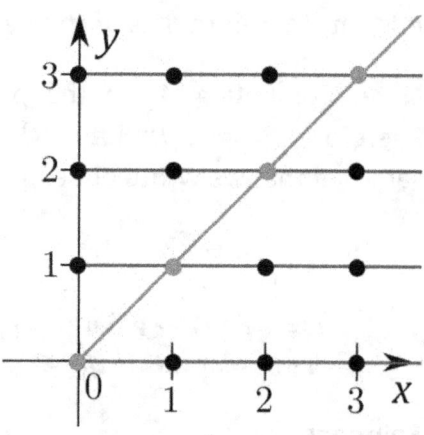

17. Let P be a partition of a set S. Prove that there is an equivalence relation $\sim$ on S for which the elements of P are the equivalence classes of $\sim$. Conversely, if $\sim$ is an equivalence relation on a set S, prove that the equivalence classes of $\sim$ form a partition of S.

Proof: Let P be a partition of S, and define the relation $\sim$ by $x \sim y$ if and only if there is $X \in P$ with $x, y \in X$.

Let $x \in S$. Since P is a partition of S, $S = \bigcup P$. So, there is $X \in P$ with $x \in X$. It follows that $x \sim x$. Therefore, $\sim$ is reflexive.

If $x \sim y$, then there is $X \in P$ with $x, y \in X$. So, $y, x \in X$ (obviously!). Thus, $y \sim x$, and therefore, $\sim$ is symmetric.

If $x \sim y$ and $y \sim z$, then there are $X, Y \in P$ with $x, y \in X$ and $y, z \in Y$. Since $y \in X$ and $y \in Y$, we have $y \in X \cap Y$. Since P is a partition and $X \cap Y \neq \emptyset$, we must have $X = Y$. So, $z \in X$. Thus, $x, z \in X$, and therefore, $x \sim z$. So, $\sim$ is transitive.

Since $\sim$ is reflexive, symmetric, and transitive on S, $\sim$ is an equivalence relation on S.

We still need to show that $P = \{[x] \mid x \in S\}$. Let $X \in P$, and let $x \in X$. We show that $X = [x]$. Let $y \in X$. Since $x, y \in X$, $x \sim y$. So $y \in [x]$. Thus, $X \subseteq [x]$. Now, let $y \in [x]$. Then $x \sim y$. So, there is $Y \in P$ such that $x, y \in Y$. Since $x \in X$ and $x \in Y$, $x \in X \cap Y$. Since P is a partition and $X \cap Y \neq \emptyset$, we must have $X = Y$. So, $y \in X$. Thus, $[x] \subseteq X$. Since $X \subseteq [x]$ and $[x] \subseteq X$, we have $X = [x]$. Since $X \in P$ was arbitrary, we have shown $P \subseteq \{[x] \mid x \in S\}$.

Now, let $X \in \{[x] \mid x \in S\}$. Then there is $x \in S$ such that $X = [x]$. Since P is a partition of S, $S = \bigcup P$. So, there is $Y \in P$ with $x \in Y$. We will show that $X = Y$. Let $y \in X$. Then $x \sim y$. So, there is $Z \in P$ with $x, y \in Z$. Since $x \in Y$ and $x \in Z$, $x \in Y \cap Z$. Since P is a partition and $Y \cap Z \neq \emptyset$, we must have $Y = Z$. So, $y \in Y$. Since $y \in X$ was arbitrary, $X \subseteq Y$. Now, let $y \in Y$. Then $x \sim y$. So, $y \in [x] = X$. Since $y \in Y$ was arbitrary, $Y \subseteq X$. Since $X \subseteq Y$ and $Y \subseteq X$, we have $X = Y$. Therefore, $X \in P$. Since $X \in \{[x] \mid x \in S\}$ was arbitrary, we have $\{[x] \mid x \in S\} \subseteq P$.

Since $P \subseteq \{[x] \mid x \in S\}$ and $\{[x] \mid x \in S\} \subseteq P$, we have $P = \{[x] \mid x \in S\}$, as desired.

Conversely, let $\sim$ be an equivalence relation on S. We first show that $\bigcup\{[x] \mid x \in S\} = S$.

Let $y \in \bigcup\{[x] \mid x \in S\}$. Then there is $x \in S$ with $y \in [x]$. By definition of $[x]$, $y \in S$. Therefore, $\bigcup\{[x] \mid x \in S\} \subseteq S$.

Now, let $y \in S$. Since $\sim$ is an equivalence relation, $y \sim y$. So, $y \in [y]$. Thus, $y \in \bigcup\{[x] \mid x \in S\}$. So, we have $S \subseteq \bigcup\{[x] \mid x \in S\}$.

Since $\bigcup\{[x] \mid x \in S\} \subseteq S$ and $S \subseteq \bigcup\{[x] \mid x \in S\}$, $\bigcup\{[x] \mid x \in S\} = S$.

We next show that if $x, y \in S$, then $[x] \cap [y] = \emptyset$ or $[x] = [y]$.

Suppose $[x] \cap [y] \neq \emptyset$ and let $z \in [x] \cap [y]$. Then $x \sim z$ and $y \sim z$. Since $\sim$ is symmetric, $z \sim y$. Since $\sim$ is transitive, $x \sim y$. Let $w \in [x]$. Then $x \sim w$. By symmetry, $y \sim x$. By transitivity, $y \sim w$. So, $w \in [y]$. Since $w \in [x]$ was arbitrary, $[x] \subseteq [y]$. By a symmetric argument, $[y] \subseteq [x]$.

Since $[x] \subseteq [y]$ and $[y] \subseteq [x]$, we have $[x] = [y]$.

Since $\bigcup\{[x] \mid x \in S\} = S$ and every pair of equivalence classes are either disjoint or equal, the set of equivalence classes partitions S. □

LEVEL 5

18. Prove that if $A \sim B$ and $C \sim D$, then $^A C \sim {}^B D$.

Proof: Suppose that $A \sim B$ and $C \sim D$. Then there exist bijections $h: A \to B$ and $k: C \to D$. Define $F: {}^A C \to {}^B D$ by $F(f)(b) = k\left(f(h^{-1}(b))\right)$.

Suppose $f, g \in {}^A C$ with $F(f) = F(g)$. Let $a \in A$ and let $b = h(a)$. We have $F(f)(b) = F(g)(b)$, or equivalently, $k\big(f(h^{-1}(b))\big) = k\big(g(h^{-1}(b))\big)$. Since k is injective, $f(h^{-1}(b)) = g(h^{-1}(b))$. Since $b = h(a)$, $a = h^{-1}(b)$. So, $f(a) = g(a)$. Since $a \in A$ was arbitrary, $f = g$. Since $f, g \in {}^A C$ were arbitrary, F is injective.

Now, let $g \in {}^B D$ and let's define $f \in {}^A C$ by $f(a) = k^{-1}\big(g(h(a))\big)$. Let $b \in B$. Then we have $F(f)(b) = k\big(f(h^{-1}(b))\big) = k\big(k^{-1}\big(g\big(h(h^{-1}(b))\big)\big)\big) = g(b)$. Since $b \in B$ was arbitrary, we have $F(f) = g$. Since $g \in {}^B D$ was arbitrary, F is surjective.

Since F is injective and surjective, ${}^A C \sim {}^B D$. $\square$

19. Prove that for any sets A, B, and C, ${}^{B \times C} A \sim {}^C({}^B A)$.

Proof: Let A, B, and C be sets, and define $F: {}^{B \times C} A \to {}^C({}^B A)$ by $F(f)(c)(b) = f(b, c)$.

Suppose $f, g \in {}^{B \times C} A$ with $F(f) = F(g)$. Let $c \in C$. Since $F(f) = F(g)$, $F(f)(c) = F(g)(c)$. So, for all $b \in B$, $F(f)(c)(b) = F(g)(c)(b)$. So, for all $b \in B$, $f(b, c) = g(b, c)$. Since $c \in C$ was arbitrary, for all $b \in B$ and $c \in C$, $f(b, c) = g(b, c)$. Therefore, $f = g$. Since $f, g \in {}^{B \times C} A$ were arbitrary, F is injective.

Let $k \in {}^C({}^B A)$ and define $f \in {}^{B \times C} A$ by $f(b, c) = k(c)(b)$. Then $F(f)(c)(b) = f(b, c) = k(c)(b)$. So, $F(f) = k$. Since $k \in {}^C({}^B A)$ was arbitrary, F is surjective.

Since F is injective and surjective, ${}^{B \times C} A \sim {}^C({}^B A)$. $\square$

20. Prove the following: (i) $\mathcal{P}(\mathbb{N}) \sim \{f \in {}^\mathbb{N}\mathbb{N} \mid f \text{ is a bijection}\}$. (ii) ${}^\mathbb{N}\mathbb{R} \nsim {}^\mathbb{R}\mathbb{N}$, given that $\mathbb{R} \sim \mathcal{P}(\mathbb{N})$.

Proofs:

(i) Let $S = \{f \in {}^\mathbb{N}\mathbb{N} \mid f \text{ is a bijection}\}$. Then $S \subseteq {}^\mathbb{N}\mathbb{N}$. So, $S \preccurlyeq {}^\mathbb{N}\mathbb{N}$ by Problem 5 (part (i)). By part (iv) of Problem 14, ${}^\mathbb{N}\mathbb{N} \sim \mathcal{P}(\mathbb{N})$. So, ${}^\mathbb{N}\mathbb{N} \preccurlyeq \mathcal{P}(\mathbb{N})$. By the transitivity of $\preccurlyeq$, $S \preccurlyeq \mathcal{P}(\mathbb{N})$.

Now, define $F: \mathcal{P}(\mathbb{N}) \to S$ by $F(A) = f_A$, where f_A is defined as follows: if $n \notin A$, then $f_A(2n) = 2n$ and $f_A(2n+1) = 2n+1$; if $n \in A$, then $f_A(2n) = 2n+1$ and $f_A(2n+1) = 2n$.

To see that F is injective, suppose that $A, B \in \mathcal{P}(\mathbb{N})$ and $A \neq B$. Without loss of generality, suppose that there is $n \in A \setminus B$. Then $f_A(2n) = 2n+1$ and $f_B(2n) = 2n$. So, $f_A \neq f_B$. Thus, $F(A) \neq F(B)$, and therefore, F is injective.

Since $S \preccurlyeq \mathcal{P}(\mathbb{N})$ and $\mathcal{P}(\mathbb{N}) \preccurlyeq S$, by the Cantor-Schroeder-Bernstein Theorem, $\mathcal{P}(\mathbb{N}) \sim S$. $\square$

(ii) We first show that $\mathbb{R} \preccurlyeq \mathcal{P}(\mathbb{Q})$. Define $f: \mathbb{R} \to \mathcal{P}(\mathbb{Q})$ by $f(x) = \{q < x \mid q \in \mathbb{Q}\}$. Let $x, y \in \mathbb{R}$ with $x \neq y$. Since trichotomy holds in $\mathbb{R}$, we have $x < y$ or $y < x$. Without loss of generality, we may assume that $x < y$. By the Density Theorem, we can choose $q \in \mathbb{Q}$ with $x < q < y$. Then $q \in f(y)$ and $q \notin f(x)$. So, $f(x) \neq f(y)$. Since $x, y \in \mathbb{R}$ were arbitrary, f is injective.

By the proof given in Problem 18, ${}^\mathbb{N}\mathbb{R} \preccurlyeq {}^\mathbb{N}\mathcal{P}(\mathbb{Q})$.

Using Problems 18 and 19, together with previous equinumerosity results, we get the following: $^{\mathbb{N}}\mathbb{R} \preccurlyeq {}^{\mathbb{N}}\mathcal{P}(\mathbb{Q}) \sim {}^{\mathbb{N}}\mathcal{P}(\mathbb{N}) \sim {}^{\mathbb{N}}({}^{\mathbb{N}}2) \sim {}^{\mathbb{N}\times\mathbb{N}}2 \sim {}^{\mathbb{N}}2 \sim \mathcal{P}(\mathbb{N}) \sim \mathbb{R} \prec \mathcal{P}(\mathbb{R}) \sim {}^{\mathbb{R}}2 \subseteq {}^{\mathbb{R}}\mathbb{N}$. It follows that $^{\mathbb{N}}\mathbb{R} \prec {}^{\mathbb{R}}\mathbb{N}$. □

Note: To help us understand the function F defined in part (i) above, let's draw a visual representation of $F(\mathbb{E})$, where $\mathbb{E}$ is the set of even natural numbers.

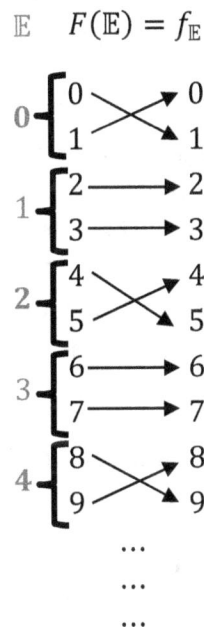

Along the left of the image we have listed the natural numbers $0, 1, 2, 3, 4, \ldots$ (we stopped at 4, but our intention is that they keep going). The elements of $\mathbb{E}$ are $0, 2, 4, \ldots$ We highlighted these in bold. We associate each natural number n with the pair $\{2n, 2n+1\}$. For example, $2 \cdot 4 = 8$ and $2 \cdot 4 + 1 = 9$. So, we associate 4 with the pair of natural numbers $\{8, 9\}$. We used left braces to indicate that association. The arrows give a visual representation of $f_{\mathbb{E}}$. Since $0 \in \mathbb{E}$, $f_{\mathbb{E}}$ swaps the corresponding pair 0 and 1. Since $1 \notin \mathbb{E}$, $f_{\mathbb{E}}$ leaves the corresponding pair 2 and 3 fixed. And so on, down the line...

The configuration of $f_{\mathbb{O}}$, where $\mathbb{O}$ is the set of odd natural numbers would be the opposite of the configuration for the evens. For example, 0 and 1 would remain fixed, while 2 and 3 would be swapped.

Problem Set 11

LEVEL 1

1. Write the elements of S_4 in cycle notation.

Solution: The elements of S_4 are $(1), (12), (13), (14), (23), (24), (34), (123), (124), (134), (234),$ $(132), (142), (143), (243), (1234), (1342), (1423), (1432), (1243), (1324), (12)(34), (13)(24),$ and $(14)(23)$.

Note: There are many ways to write the same permutation in cycle notation. For example, (12) is the same permutation as (21). Both permutations send 1 to 2, 2 to 1, and the other numbers to themselves. As a more extreme example, (123) is the same permutation as $(13)(12)$. The latter way of writing the permutation is a bit less "natural" because the number 1 is repeated twice. It turns out that every permutation can be written as a product of cycles in a way that each number appears no more than once. In the solution above, I have written each permutation in the most "natural" way.

2. Draw a group multiplication table for S_3. Let $H = \{(1), (123), (132)\}$ and $K = \{(1), (12)\}$. Show that H and K are subgroups of S_3 and determine which of these is a normal subgroup of S_3.

Solution:

$(S_3, \circ)$	(1)	(12)	(13)	(23)	(123)	(132)
(1)	(1)	(12)	(13)	(23)	(123)	(132)
(12)	(12)	(1)	(132)	(123)	(23)	(13)
(13)	(13)	(123)	(1)	(132)	(12)	(23)
(23)	(23)	(132)	(123)	(1)	(13)	(12)
(123)	(123)	(13)	(23)	(12)	(132)	(1)
(132)	(132)	(23)	(12)	(13)	(1)	(123)

We now show that H and K are subgroups of S_3. The restricted tables look as follows:

$(H, \circ)$	(1)	(123)	(132)
(1)	(1)	(123)	(132)
(123)	(123)	(132)	(1)
(132)	(132)	(1)	(123)

$(K, \circ)$	(1)	(12)
(1)	(1)	(12)
(12)	(12)	(1)

From the tables, we can see that each of H and K are closed under $\circ$. Also, each of H and K contain the identity (1). In H, (123) and (132) are inverses of each other, and in K, (12) is its own inverse. It follows that both H and K are subgroups of S_3.

Now K is **not** normal in S_3 because $(13)(12)(13)^{-1} = (13)(12)(13) = (23)$ and $(23) \notin K$.

Finally, we show that $H \triangleleft S_3$. We do this by brute force:

$$(12)(123)(12) = (132), (13)(123)(13) = (132), (23)(123)(23) = (132),$$
$$(12)(132)(12) = (123), (13)(132)(13) = (123), (23)(132)(23) = (123).$$

We do not need to check conjugation by the elements in H because we already know that H is a subgroup of S_3. We see that for all $g \in S_3$ and $h \in H$, we have $ghg^{-1} \in H$, and therefore, $H \triangleleft S_3$.

LEVEL 2

3. A **Gaussian integer** is a complex number of the form $a + bi$, where $a, b \in \mathbb{Z}$. Let $\mathbb{Z}[i]$ be the set of Gaussian integers. Prove that $(\mathbb{Z}[i], +, \cdot)$ is a subring of $(\mathbb{C}, +, \cdot)$.

Proof: Let $a + bi, c + di \in \mathbb{Z}[i]$. Then $(a + bi) + (c + di) = (a + c) + (b + d)i$. Since $\mathbb{Z}$ is closed under addition, $a + c \in \mathbb{Z}$ and $b + d \in \mathbb{Z}$. Therefore, $(a + bi) + (c + di) \in \mathbb{Z}[i]$, and so, $\mathbb{Z}[i]$ is closed under addition. Also, $(a + bi)(c + di) = (ac - bd) + (ad + bc)i$. Since $\mathbb{Z}$ is closed under multiplication, $ac, bd, ad, bc \in \mathbb{Z}$. Since $\mathbb{Z}$ has the additive inverse property, $-(bd) \in \mathbb{Z}$. Since $\mathbb{Z}$ is closed under addition, we have $ad + bc \in \mathbb{Z}$ and $ac - bd = ac + (-(bd)) \in \mathbb{Z}$. Therefore, $(a + bi)(c + di) \in \mathbb{Z}[i]$, and so, $\mathbb{Z}[i]$ is closed under multiplication. The additive inverse of $a + bi$ is $-a - bi = -a + (-b)i$. Since $\mathbb{Z}$ has the additive inverse property, $-a, -b \in \mathbb{Z}$, and so, $-a - bi \in \mathbb{Z}[i]$. Finally, $1 = 1 + 0i \in \mathbb{Z}[i]$ because $1, 0 \in \mathbb{Z}$. It follows that $(\mathbb{Z}[i], +, \cdot)$ is a subring of $(\mathbb{C}, +, \cdot)$. $\square$

4. Let $(G, \star)$ be a group with H a nonempty subset of G. Prove that $(H, \star)$ is a subgroup of $(G, \star)$ if and only if for all $g, h \in H$, $g \star h^{-1} \in H$.

Proofs: Let $(G, \star)$ be a group with H a nonempty subset of G.

First, let $(H, \star)$ be a subgroup of $(G, \star)$, and let $g, h \in H$. Since H has the inverse property, h^{-1} exists in H. Since H is closed under $\star$, $g \star h^{-1} \in H$. Since $g, h \in H$ were arbitrary, we have shown that for all $g, h \in H$, $g \star h^{-1} \in H$.

Conversely, suppose that for all $g, h \in H$, $g \star h^{-1} \in H$. Since $H \neq \emptyset$, there is $g \in H$. Therefore, $e = g \star g^{-1} \in H$. Now, let $h \in H$. Since $e \in H$, $h^{-1} = e \star h^{-1} \in H$. Finally, let $g, h \in H$. We just showed that $h^{-1} \in H$. Therefore, $g \star h = g \star (h^{-1})^{-1} \in H$. So, H is closed under $\star$. Thus, $(H, \star)$ is a subgroup of $(G, \star)$. $\square$

5. Let $(R, +, \cdot)$ be a ring and define addition and multiplication on $R \times R$ componentwise, as was done in part 4 of Example 11.6. Prove that $(R \times R, +, \cdot)$ is a ring and that $(R, +, \cdot)$ is isomorphic to a subring of $(R \times R, +, \cdot)$.

Proof: Let $(a, b), (c, d) \in R \times R$. Then $a, b, c, d \in R$. Since R is closed under addition and multiplication, $a + c, b + d, ac, bd \in R$. So, $(a + c, b + d), (ac, bd) \in R \times R$. So, $R \times R$ is closed under addition and multiplication.

Let $(a, b), (c, d), (e, f) \in R \times R$. Since addition and multiplication are associative in R, we have

$$(a,b) + ((c,d) + (e,f)) = (a,b) + (c+e, d+f) = (a + (c+e), b + (d+f))$$
$$= ((a+c)+e, (b+d)+f) = (a+c, b+d) + (e,f) = ((a,b) + (c,d)) + (e,f).$$
$$(a,b) \cdot ((c,d) \cdot (e,f)) = (a,b) \cdot (ce, df) = (a(ce), b(df))$$
$$= ((ac)e, (bd)f) = (ac, bd) \cdot (e,f) = ((a,b) \cdot (c,d)) \cdot (e,f).$$

So, addition and multiplication are associative in $R \times R$.

For all $(a,b) \in R \times R$, we have
$$(0,0) + (a,b) = (0+a, 0+b) = (a,b) \text{ and } (a,b) + (0,0) = (a+0, b+0) = (a,b).$$
$$(1,1) \cdot (a,b) = (1a, 1b) = (a,b) \text{ and } (a,b) \cdot (1,1) = (a \cdot 1, b \cdot 1) = (a,b).$$

So, $(0,0)$ is an additive identity and $(1,1)$ is a multiplicative identity.

For all $(a,b) \in R \times R$, we have
$$(a,b) + (-a,-b) = (a-a, b-b) = (0,0) \text{ and } (-a,-b) + (a,b) = (-a+a, -b+b) = (0,0).$$

So, $(-a, -b)$ is an additive inverse of (a,b).

Let $(a,b), (c,d) \in R \times R$. Since addition is commutative in R, we have
$$(a,b) + (c,d) = (a+c, b+d) = (c+a, d+b) = (c,d) + (a,b).$$

So, addition is commutative in $R \times R$.

Let $(a,b), (c,d), (e,f) \in R \times R$. Since multiplication is distributive over addition in R, we have
$$(a,b)((c,d) + (e,f)) = (a,b)(c+e, d+f) = (a(c+e), b(d+f))$$
$$= (ac+ae, bd+bf) = (ac, bd) + (ae, bf) = (a,b)(c,d) + (a,b)(e,f).$$

So, multiplication is distributive over addition in $R \times R$.

Therefore, $(R \times R, +, \cdot)$ is ring.

Let $S = \{(x,x) \mid x \in R\}$. Then $S \subseteq R \times R$, $(0,0) \in S$, and $(1,1) \in S$. If $(x,x), (y,y) \in S$, then we have $(x,x) - (y,y) = (x-y, x-y) \in S$ and $(x,x) \cdot (y,y) = (xy, xy) \in S$. So, S is a subring of $R \times R$ (note that we used Problem 4 here).

Define $f: R \to S$ by $f(x) = (x,x)$. Clearly, f is bijective. If $x, y \in R$, we have
$$f(x+y) = (x+y, x+y) = (x,x) + (y,y) = f(x) + f(y).$$
$$f(xy) = (xy, xy) = (x,x) \cdot (y,y) = f(x) \cdot f(y).$$

So, f is a homomorphism. Therefore, $f: R \cong S$. $\square$

LEVEL 3

6. Prove that there are exactly two ring homomorphisms from $\mathbb{Z}$ to itself.

Proof: The identity function $i_\mathbb{Z}: \mathbb{Z} \to \mathbb{Z}$ and the zero function $0: \mathbb{Z} \to \mathbb{Z}$ are both ring homomorphisms from $\mathbb{Z}$ to itself. We show that these are the only ones.

Let $f: \mathbb{Z} \to \mathbb{Z}$ be a ring homomorphism and suppose that $f(1) = n$. Then we also have
$$f(1) = f(1 \cdot 1) = f(1) \cdot f(1) = n \cdot n = n^2.$$

So, $n^2 = n$. Therefore, $n^2 - n = 0$, and so, $n(n - 1) = 0$. It follows that $n = 0$ or $n = 1$.

Now, it is easy to show that $f(k) = k \cdot f(1)$ for all $k \in \mathbb{Z}$. For $k \in \mathbb{N}$, use the principle of mathematical induction. Then if $k < 0$, we have $f(k) = f(-(-k)) = -f(-k) = -(-k)f(1) = kf(1)$. $f(0) = 0$ follows from Theorem 11.1.

So, if $f(1) = 0$, then for all $k \in \mathbb{Z}$, $f(k) = kf(1) = k \cdot 0 = 0$. In this case, f is the zero function.

If $f(1) = 1$, then for all $k \in \mathbb{Z}$, $f(k) = kf(1) = k \cdot 1 = k$. In this case, f is the identity function. □

7. Prove the following: (i) Ring isomorphism is an equivalence relation. (ii) If we let $\text{Aut}(R)$ be the set of automorphisms of a ring R, then $(\text{Aut}(R), \circ)$ is a group, where $\circ$ is composition.

Proofs:

(i) $i_R: R \to R$ is a bijection and if $x, y \in R$, then $i_R(x + y) = x + y = i_R(x) + i_R(y)$ and $i_R(xy) = xy = i_R(x)i_R(y)$. Also, $i_R(1_R) = 1_R$. So, i_R is an isomorphism from R to itself. Therefore, $\cong$ is reflexive.

Suppose that $f: R \to S$ is an isomorphism from R to S. We already know that $f^{-1}: S \to R$ is a bijection from S to R. Let $x, y \in S$ and let $z, w \in R$ with $f(z) = x$ and $f(w) = y$. Then $f(z + w) = f(z) + f(w) = x + y$, and so, $f^{-1}(x + y) = z + w = f^{-1}(x) + f^{-1}(y)$. Also, $f(zw) = f(z)f(w) = xy$, so that $f^{-1}(xy) = zw = f^{-1}(x)f^{-1}(y)$. Finally, $f^{-1}(1_S) = 1_R$. Therefore, f^{-1} is an isomorphism from S to R, and so, $\cong$ is symmetric.

Suppose that $f: R \to S, g: S \to T$ are isomorphisms. We already know that $g \circ f: R \to T$ is a bijection. If $x, y \in R$, then we have
$$(g \circ f)(x + y) = g(f(x + y)) = g(f(x) + f(y))$$
$$= g(f(x)) + g(f(y)) = (g \circ f)(x) + (g \circ f)(y)$$
and
$$(g \circ f)(xy) = g(f(xy)) = g(f(x) \cdot f(y))$$
$$= g(f(x)) \cdot g(f(y)) = (g \circ f)(x) \cdot (g \circ f)(y).$$

Also, $(g \circ f)(1_R) = g(f(1_R)) = g(1_S) = 1_T$. So, $g \circ f$ is an isomorphism from R to T, and so, $\cong$ is transitive.

Since $\cong$ is reflexive, symmetric, and transitive, $\cong$ is an equivalence relation. □

(ii) Let R be a ring and let $f, g \in \text{Aut}(R)$. Then $g \circ f$ is an automorphism by the proof of transitivity from part (i). So $\text{Aut}(R)$ is closed under $\circ$. We proved that $\circ$ is associative in $S(R)$ in part 5 of Example 11.8. Since $\text{Aut}(R) \subseteq S(R)$, $\circ$ is associative in $\text{Aut}(R)$ as well. We proved that i_R satisfies $i_R \circ f = f$ and $f \circ i_R = f$ for all $f \in \text{Aut}(R)$ in the same example and we proved i_R is an automorphism in part (i) above. Let $f \in \text{Aut}(R)$. By Theorem 10.6, we have $f^{-1} \circ f = f \circ f^{-1} = i_R$, and we proved that f^{-1} is an automorphism in part (i) above. It follows that $(\text{Aut}(R), \circ)$ is a group. □

8. Let G be a group with H and K subgroups of G, and let $G = H \cup K$. Prove that $H = G$ or $K = G$.

Proof: Let G be a group, let H and K be subgroups of G, and let $G = H \cup K$. Suppose toward contradiction that $H \neq G$ and $K \neq G$. Then there exist $a \in G \setminus H$ and $b \in G \setminus K$. Since $G = H \cup K$, we have $a \in K$ and $b \in H$. Since G is a group, $ab \in G$. So, $ab \in H$ or $ab \in K$. Without loss of generality, let $ab \in H$. Since $b \in H$ and H is a group, $b^{-1} \in H$, and thus, $a = ae = a(bb^{-1}) = (ab)b^{-1} \in H$. This contradicts our assumption that $a \in G \setminus H$. So, we must have $H = G$ or $K = G$. □

9. Prove that a commutative ring R is a field if and only if the only ideals of R are $\{0\}$ and R.

Proof: Let R be a commutative ring. First assume that R is a field and let I be an ideal of R such that $I \neq \{0\}$. Then there is $a \in I$ with $a \neq 0$. Since R is a field and $a \neq 0$, a^{-1} exists. Since I is an ideal, we have $1 = aa^{-1} \in I$. Now, let $x \in R$ be arbitrary. Then $x = x \cdot 1 \in I$. Therefore, $R \subseteq I$. By definition of an ideal, $I \subseteq R$. So, $I = R$. Since I was an arbitrary ideal not equal to $\{0\}$, the only ideals of R are $\{0\}$ and R.

Conversely, assume that the only ideals of R are $\{0\}$ and R. Let $a \in R$ with $a \neq 0$, and let $I = \{ax \mid x \in R\}$. We show that I is an ideal of R. First, if $ax, ay \in I$, then $ax + ay = a(x + y) \in I$. So, I is closed under addition. Since $a \cdot 0 = 0$, $0 \in I$. If $ax \in I$, then $-(ax) = x(-a) \in I$ (Why?). It follows that $(I, +)$ is a subgroup of $(R, +)$. If $ax \in I$ and $b \in R$, then $(ax)b = a(xb) = a(bx) = (ab)x \in I$. So, I absorbs R. Therefore, I is an ideal of R.

Since $a = ae \in I$ and $a \neq 0$, $I \neq \{0\}$. By assumption, we must have $I = R$. Since $1 \in R$, $1 \in I$. Therefore, there is $b \in R$ such that $ab = 1$. So, $b = a^{-1}$. Since $a \neq 0$ was arbitrary, we have shown that R has the multiplicative inverse property, and therefore, R is a field. □

Note: The ideal $I = \{ax \mid x \in R\}$ is called the **principal ideal generated by a**.

10. Prove that if X is a nonempty set of normal subgroups of a group G then $\cap X$ is a normal subgroup of G. Similarly, prove that if X is a nonempty set of ideals of a ring R, then $\cap X$ is an ideal of R. Is the union of normal subgroups always a normal subgroup? Is the union of ideals always an ideal?

Proofs: Let X be a nonempty set of normal subgroups of a group G. Since for all $H \in X$, $e \in H$, $e \in \cap X$, and so, $\cap X \neq \emptyset$. Let $g, h \in \cap X$. Then for all $H \in X$, $g, h \in H$. By Problem 4, for all $H \in X$, $gh^{-1} \in H$. So, $gh^{-1} \in \cap X$. Again, by Problem 4, $\cap X$ is a subgroup of G. Now, let $h \in \cap X$, and let $g \in G$. Then for all $H \in X$, $h \in H$, and since each H is a normal subgroup of G, $ghg^{-1} \in H$. So, $ghg^{-1} \in \cap X$. Therefore, $\cap X$ is a normal subgroup of G. □

Let X be a nonempty set of ideals of a ring R. Since for all $I \in X$, $0 \in I$, $0 \in \cap X$, and so, $\cap X \neq \emptyset$. Let $x, y \in \cap X$. Then for all $I \in X$, $x, y \in I$. By Problem 4, for all $I \in X$, $x - y \in I$. So, $x - y \in \cap X$. Again, by Problem 4, $(\cap X, +)$ is a subgroup of $(R, +)$. Now, let $x \in \cap X$, and let $y \in R$. Then for all $I \in X$, $x \in I$, and since each I is an ideal of R, $xy \in I$. So, $xy \in \cap X$. Therefore, $\cap X$ absorbs R, and so, $\cap X$ is an ideal of R. □

Since $(\mathbb{Z}, +)$ is a commutative group, all subgroups are normal. In particular $(2\mathbb{Z}, +)$ and $(3\mathbb{Z}, +)$ are normal subgroups of $(\mathbb{Z}, +)$. Now, $2, 3 \in (2\mathbb{Z}, +) \cup (3\mathbb{Z}, +)$, but $2 + 3 = 5 \notin (2\mathbb{Z}, +) \cup (3\mathbb{Z}, +)$. So, $(2\mathbb{Z}, +) \cup (3\mathbb{Z}, +)$ is not closed under addition and is therefore not a subgroup of $(\mathbb{Z}, +)$.

Since $(2\mathbb{Z}, +, \cdot)$ and $(3\mathbb{Z}, +, \cdot)$ are ideals of $(\mathbb{Z}, +, \cdot)$, the same argument in the last paragraph shows that the union of ideals is **not** always an ideal.

11. Let $\mathbb{Z}_n[x] = \{a_n x^n + a_{n-1} x^{n-1} + \cdots + a_1 x + a_0 \mid a_0, a_1, \ldots, a_n \in \mathbb{Z}\}$. In other words, $\mathbb{Z}_n[x]$ consists of all polynomials of degree at most n. Prove that $(\mathbb{Z}_n[x], +)$ is a commutative group for $n = 0, 1$, and 2, where addition is defined in the "usual way." Then prove that $\mathbb{Z}_0[x]$ is a subgroup of $\mathbb{Z}_1[x]$ and $\mathbb{Z}_1[x]$ is a subgroup of $\mathbb{Z}_2[x]$. What if we replace "all polynomials of degree at most n" with "all polynomials of degree n?"

Proof: $\mathbb{Z}_0[x] = \{a_0 \mid a_0 \in \mathbb{Z}\} = \mathbb{Z}$, and we already know that $(\mathbb{Z}, +)$ is a group.

$\mathbb{Z}_1[x] = \{a_1 x + a_0 \mid a_0, a_1 \in \mathbb{Z}\}$. Let $a_1 x + a_0, b_1 x + b_0 \in \mathbb{Z}_1[x]$. We have
$$(a_1 x + a_0) + (b_1 x + b_0) = (a_1 + b_1) x + (a_0 + b_0).$$

Since $\mathbb{Z}$ is closed under addition, $a_1 + b_1, a_0 + b_0 \in \mathbb{Z}$. Therefore, $\mathbb{Z}_1[x]$ is closed under addition.

Simple computations show that addition is associative in $\mathbb{Z}_1[x]$, $0 = 0x + 0$ is an additive identity in $\mathbb{Z}_1[x]$, and the additive inverse of $a_1 x + a_0$ is $-a_1 x - a_0$. One more simple computation can be used to verify that addition is commutative in $\mathbb{Z}_1[x]$. So, $(\mathbb{Z}_1[x], +)$ is a commutative group.

$\mathbb{Z}_2[x] = \{a_2 x^2 + a_1 x + a_0 \mid a_0, a_1, a_2 \in \mathbb{Z}\}$. Let $a_2 x^2 + a_1 x + a_0, b_2 x^2 + b_1 x + b_0 \in \mathbb{Z}_1[x]$. We have
$$(a_2 x^2 + a_1 x + a_0) + (b_2 x^2 + b_1 x + b_0) = (a_2 + b_2) x^2 + (a_1 + b_1) x + (a_0 + b_0).$$

Since $\mathbb{Z}$ is closed under addition, $a_2 + b_2, a_1 + b_1, a_0 + b_0 \in \mathbb{Z}$. Therefore, $\mathbb{Z}_2[x]$ is closed under addition.

Simple computations show that addition is associative in $\mathbb{Z}_2[x]$, $0 = 0x^2 + 0x + 0$ is an additive identity in $\mathbb{Z}_2[x]$, and the additive inverse of $a_2 x^2 + a_1 x + a_0$ is $-a_2 x^2 - a_1 x - a_0$. One more simple computation can be used to verify that addition is commutative in $\mathbb{Z}_2[x]$. So, $(\mathbb{Z}_2[x], +)$ is a commutative group.

Let $a_0 \in \mathbb{Z}_0[x] = \mathbb{Z}$. Then $a_0 = 0x + a_0 \in \mathbb{Z}_1[x]$. So, $\mathbb{Z}_0[x] \subseteq \mathbb{Z}_1[x]$. Since $\mathbb{Z}_0[x]$ is a group under addition, $\mathbb{Z}_0[x] \leq \mathbb{Z}_1[x]$. Similarly, if $a_1 x + a_0 \in \mathbb{Z}_1[x]$, then $a_1 x + a_0 = 0x^2 + a_1 x + a_0 \in \mathbb{Z}_2[x]$. Therefore, $\mathbb{Z}_1[x] \subseteq \mathbb{Z}_2[x]$. Since we showed that $\mathbb{Z}_1[x]$ is a group under addition, $\mathbb{Z}_1[x] \leq \mathbb{Z}_2[x]$.

Let A be the set of polynomials of degree 1. Then x and $-x$ are in A, but $x + (-x) = 0 \notin A$. So, A is not closed under addition, and therefore, $(A, +)$ is **not** a group. □

LEVEL 4

12. Let N be a normal subgroup of a group G. For each $g \in G$, let $gN = \{gx \mid x \in N\}$. Prove that $gN = hN$ if and only if $gh^{-1} \in N$. Let $G/N = \{gN \mid g \in G\}$. Prove that $(G/N, \circ)$ is a group, where $\circ$ is defined by $gN \circ hN = (gh)N$.

Proof: Let $N \triangleleft G$, and let $g, h \in G$. Suppose that $gN = hN$. Since $N \leq G$, $e \in N$. So, $g = ge \in gN$. Since $gN = hN$, $g \in hN$. So, there is $x \in N$ with $g = hx$. Then since $N \triangleleft G$, $gh^{-1} = hxh^{-1} \in N$.

Conversely, assume that $gh^{-1} \in N$ and let $x \in gN$. Then there is $y \in N$ with $x = gy$, or equivalently, $y = g^{-1}x$. So, $g^{-1}x = y \in N$. Since $N \triangleleft G$, $(hg^{-1})(xh^{-1}) = h(g^{-1}x)h^{-1} \in N$. So, there is $z \in N$ with $(gh^{-1})^{-1}(xh^{-1}) = h(g^{-1}x)h^{-1} = z$. Thus, $xh^{-1} = (gh^{-1})z \in N$. Again, since $N \triangleleft G$, it follows that $h^{-1}x = (h^{-1}x)(h^{-1}h) = h^{-1}(xh^{-1})h \in N$. So, there is $w \in N$ with $h^{-1}x = w$. So, $x = hw$. Since $w \in N$, $x \in hN$. Since $x \in gN$ was arbitrary, $gN \subseteq hN$. By a symmetric argument, $hN \subseteq gN$. Therefore, $gN = hN$.

Now, let $G/N = \{gN \mid g \in G\}$ and define $\circ$ by $gN \circ hN = (gh)N$.

We first show that $\circ$ is well defined. Suppose $gN = jN$ and $hN = kN$. By the theorem we just proved, we have $gj^{-1} \in N$ and $hk^{-1} \in N$. Since $N \triangleleft G$ and $hk^{-1} \in N$, we have $j(hk^{-1})j^{-1} \in N$. So, $(gh)(jk)^{-1} = ghk^{-1}j^{-1} = (gj^{-1})(jhk^{-1}j^{-1}) \in N$. Again, by the theorem we just proved $(gh)N = (jk)N$.

Closure and associativity are clear.

Note that $eN = N$ because for all $g \in G$, $g = eg$. It follows that $gN \circ N = gN \circ eN = (ge)N = gN$ and $N \circ gN = eN \circ gN = (eg)N = gN$. Therefore, N is the identity element of G/N.

Finally, $(gN)^{-1} = g^{-1}N$ because $gN \circ g^{-1}N = (gg^{-1})N = eN = N$. □

13. Let I be an ideal of a ring R. For each $x \in R$, let $x + I = \{x + z \mid z \in I\}$. Prove that $x + I = y + I$ if and only if $x - y \in I$. Let $R/I = \{x + I \mid x \in R\}$. Prove that $(R/I, +, \cdot)$ is a ring, where addition and multiplication are defined by $(x + I) + (y + I) = (x + y) + I$ and $(x + I)(y + I) = xy + I$.

Proof: Let $I \triangleleft R$. Then $(I, +)$ is a commutative subgroup of $(R, +)$. All subgroups of an abelian group are normal subgroups, and so, by Problem 12, $x + I = y + I$ if and only if $x - y \in I$ (note that we're using additive notation because $(I, +)$ is an additive group).

Let $R/I = \{x + I \mid x \in R\}$ and define addition and multiplication by $(x + I) + (y + I) = (x + y) + I$ and $(x + I)(y + I) = xy + I$. By Problem 12, $(R/I, +)$ is a group. Commutativity of addition is clear.

We next check that multiplication is well defined. Suppose $x + I = z + I$ and $y + I = w + I$. By the theorem we just proved, we have $x - z \in I$ and $y - w \in I$. Since $I \triangleleft R$, we have $(x - z)w \in I$ and $x(y - w) \in I$. Since $(I, +)$ is a group, we have

$$xy - zw = xy - xw + xw - zw = x(y - w) + (x - z)w \in I$$

By the theorem we just proved, $xy + I = zw + I$.

Associativity of multiplication is clear.

Finally, we show that $1 + I$ is the multiplicative identity of R/I. For all $x \in R$, $1x = x$. It follows that $(1 + I) \cdot (x + I) = 1x + I = x + I$. □

> 14. Let $\mathbb{Z}_n = \{[k] \mid k \in \mathbb{Z}\}$, where $[k]$ is the equivalence class of k under the equivalence $\equiv_n$. Prove that $(\mathbb{Z}_n, +, \cdot)$ is a ring, where addition and multiplication are defined by $[x] + [y] = [x + y]$ and $[xy] = [x] \cdot [y]$. Then prove that $\mathbb{Z}/n\mathbb{Z} \cong \mathbb{Z}_n$. Find the ideals of $\mathbb{Z}/15\mathbb{Z}$ and $\mathbb{Z}_{15}$ and show that there is a natural one-to-one correspondence between them.

Proof: We first show that addition is well defined. Suppose $[x] = [z]$ and $[y] = [w]$. Then $x \equiv_n z$ and $y \equiv_n w$, so that $n \mid x - z$ and $n \mid y - w$. Therefore, there are $j, k \in \mathbb{Z}$ such that $x - z = nj$ and $y - w = nk$. It follows that $(x + y) - (z + w) = (x - z) + (y - w) = nj + nk = n(j + k)$. Since $\mathbb{Z}$ is closed under addition, $j + k \in \mathbb{Z}$, so that $n \mid (x + y) - (z + w)$. Thus, $x + y \equiv_n z + w$, and so, $[x + y] = [z + w]$.

Let $[x], [y] \in \mathbb{Z}_n$. Then $x, y \in \mathbb{Z}$. Since $\mathbb{Z}$ is closed under addition, $x + y \in \mathbb{Z}$. Thus, $[x + y] \in \mathbb{Z}_n$. So, $\mathbb{Z}_n$ is closed under addition.

Let $[x], [y], [z] \in \mathbb{Z}_n$. Since addition is associative in $\mathbb{Z}$, we have

$$([x] + [y]) + [z] = [x + y] + [z] = [(x + y) + z]$$
$$= [x + (y + z)] = [x] + [y + z] = [x] + ([y] + [z]).$$

So, addition is associative in $\mathbb{Z}_n$.

If $[x] \in \mathbb{Z}_n$, then $[0] + [x] = [0 + x] = [x]$ and $[x] + [0] = [x + 0] = [x]$. So, $[0]$ is an identity element for addition in $\mathbb{Z}_n$.

If $[x] \in \mathbb{Z}_n$, then $[-x] + [x] = [-x + x] = [0]$ and $[x] + [-x] = [x + (-x)] = [0]$. So, $-[x] = [-x]$. Therefore, the inverse property holds for addition in $\mathbb{Z}_n$.

Let $[x], [y] \in \mathbb{Z}_n$. Since addition is commutative in $\mathbb{Z}$, $[x] + [y] = [x + y] = [y + x] = [y] + [x]$. So, addition is commutative in $\mathbb{Z}_n$.

We next show that multiplication is well defined. Suppose $[x] = [z]$ and $[y] = [w]$. As in the proof for addition, there are $j, k \in \mathbb{Z}$ such that $x - z = nj$ and $y - w = nk$. It follows that

$$xy - zw = xy - zy + zy - zw = (x - z)y + z(y - w) = (nj)y + z(nk) = n(jy + zk).$$

Since $\mathbb{Z}$ is closed under addition and multiplication, $jy + zk \in \mathbb{Z}$, and so, $n|xy - zw$. Therefore, $xy \equiv_n zw$, and so, $[xy] = [yw]$.

Let $[x], [y], [z] \in \mathbb{Z}_n$. Since multiplication is associative in $\mathbb{Z}$, we have

$$([x] \cdot [y]) \cdot [z] = [xy] \cdot [z] = [(xy)z] = [x(yz)] = [x] \cdot [yz] = [x] \cdot ([y] \cdot [z])$$

So, multiplication is associative in $\mathbb{Z}_n$.

If $[x] \in \mathbb{Z}_n$, then $[1] \cdot [x] = [1x] = [x]$ and $[x] \cdot [1] = [x \cdot 1] = [x]$. So, $[1]$ is an identity element for multiplication in $\mathbb{Z}_n$.

Define $f: \mathbb{Z}/n\mathbb{Z} \to \mathbb{Z}_n$ by $f(x + n\mathbb{Z}) = [x]$. Suppose that $f(x + n\mathbb{Z}) = f(y + n\mathbb{Z})$, so that $[x] = [y]$. Then $x \equiv_n y$. So, $n|x - y$, and therefore, there is $k \in \mathbb{Z}$ such that $x - y = nk$. So, $x - y \in n\mathbb{Z}$, and so, by Problem 13, $x + n\mathbb{Z} = y + n\mathbb{Z}$. Since $x + n\mathbb{Z}, y + n\mathbb{Z} \in \mathbb{Z}/n\mathbb{Z}$ were arbitrary, f is injective.

f is surjective because if $[x] \in \mathbb{Z}_n$, then $x \in \mathbb{Z}$ and $f(x + n\mathbb{Z}) = [x]$.

Now, we have

$$f\big((x + n\mathbb{Z}) + (y + n\mathbb{Z})\big) = f\big((x + y) + n\mathbb{Z}\big) = [x + y] = [x] + [y] = f(x + n\mathbb{Z}) + f(y + n\mathbb{Z})$$

$$f\big((x + n\mathbb{Z})(y + n\mathbb{Z})\big) = f\big((xy) + n\mathbb{Z}\big) = [xy] = [x] \cdot [y] = f(x + n\mathbb{Z}) \cdot f(y + n\mathbb{Z})$$

Therefore, $f: \mathbb{Z}/n\mathbb{Z} \cong \mathbb{Z}_n$.

$\mathbb{Z}/15\mathbb{Z} = \{15\mathbb{Z}, 1 + 15\mathbb{Z}, \ldots, 14 + 15\mathbb{Z}\}$ and $\mathbb{Z}_{15} = \{[0], [1], \ldots, [14]\}$.

The ideals of $\mathbb{Z}/15\mathbb{Z}$ are $\mathbb{Z}/15\mathbb{Z}, 3\mathbb{Z}/15\mathbb{Z}, 5\mathbb{Z}/15\mathbb{Z}, 15\mathbb{Z}/15\mathbb{Z}$ and these correspond to the ideals of $\mathbb{Z}_{15}$, which are $\mathbb{Z}_{15}, \{[0], [3], [6], [9], [12]\}, \{[0], [5], [10]\}$, and $\{[0]\}$. □

LEVEL 5

15. Let $\mathbb{Z}[x] = \{a_k x^k + a_{k-1} x^{k-1} + \cdots + a_1 x + a_0 \mid k \in \mathbb{N} \land a_0, a_1, \ldots, a_k \in \mathbb{Z}\}$. $(\mathbb{Z}[x], +, \cdot)$ with addition and multiplication defined in the "usual way" is called the **polynomial ring over** $\mathbb{Z}$. Prove that $(\mathbb{Z}[x], +, \cdot)$ is a ring. Then prove that $(\mathbb{Z}_n[x], +, \cdot)$ is **not** a subring of $(\mathbb{Z}[x], +, \cdot)$ for any $n \in \mathbb{N}$. Let $R[x] = \{a_k x^k + a_{k-1} x^{k-1} + \cdots + a_1 x + a_0 \mid k \in \mathbb{N} \land a_0, a_1, \ldots, a_k \in R\}$ for an arbitrary ring R. Is $(R[x], +, \cdot)$ a ring?

Proof: Addition in $(\mathbb{Z}[x], +, \cdot)$ is defined by

$$a_k x^k + a_{k-1} x^{k-1} + \cdots + a_1 x + a_0) + (b_k x^k + b_{k-1} x^{k-1} + \cdots + b_1 x + b_0)$$
$$= (a_k + b_k) x^k + (a_{k-1} + b_{k-1}) x^{k-1} + \cdots + (a_1 + b_1) x + (a_0 + b_0).$$

Since $\mathbb{Z}$ is closed under addition, $a_k + b_k, a_{k-1} + b_{k-1}, \ldots, a_1 + b_1, a_0 + b_0 \in \mathbb{Z}$. Therefore, $\mathbb{Z}[x]$ is closed under addition.

Simple computations show that addition is associative in $\mathbb{Z}[x]$, $0 = 0x^k + 0x^{k-1} + \cdots + 0x + 0$ is an additive identity in $\mathbb{Z}[x]$, and the additive inverse of $a_k x^k + a_{k-1} x^{k-1} + \cdots + a_1 x + a_0$ is $-a_k x^k - a_{k-1} x^{k-1} - \cdots - a_1 x - a_0$. One more simple computation can be used to verify that addition is commutative.

Here comes the hard part... Multiplication in $(\mathbb{Z}[x], +, \cdot)$ is defined by

$$(a_k x^k + a_{k-1} x^{k-1} + \cdots + a_1 x + a_0)(b_k x^k + b_{k-1} x^{k-1} + \cdots + b_1 x + b_0)$$
$$= c_k x^k + c_{k-1} x^{k-1} + c_1 x + c_0,$$

where $c_i = a_i b_0 + a_{i-1} b_1 + \cdots + a_1 b_{i-1} + a_0 b_i$ for each $i = 0, 1, \ldots, k$.

Since $\mathbb{Z}$ is closed under addition and multiplication, $c_i = a_i b_0 + a_{i-1} b_1 + \cdots + a_1 b_{i-1} + a_0 b_i \in \mathbb{Z}$ for each $i = 0, 1, \ldots, k$. Therefore, $\mathbb{Z}[x]$ is closed under multiplication.

We show that 1 is a multiplicative identity for $\mathbb{Z}[x]$. We have

$$1 \cdot (a_k x^k + a_{k-1} x^{k-1} + \cdots + a_1 x + a_0) = c_k x^k + c_{k-1} x^{k-1} + c_1 x + c_0,$$

where $c_i = 0 a_0 + 0 a_1 + \cdots + 0 a_{i-1} + 1 a_i = a_i$. So,

$$1 \cdot (a_k x^k + a_{k-1} x^{k-1} + \cdots + a_1 x + a_0) = a_k x^k + a_{k-1} x^{k-1} + \cdots + a_1 x + a_0.$$

Similarly, $(a_k x^k + a_{k-1} x^{k-1} + \cdots + a_1 x + a_0) \cdot 1 = c_k x^k + c_{k-1} x^{k-1} + c_1 x + c_0,$

where $c_i = a_i \cdot 1 + a_{i-1} \cdot 0 + \cdots + a_1 \cdot 0 + a_0 \cdot 0 = a_i$. So,

$$(a_k x^k + a_{k-1} x^{k-1} + \cdots + a_1 x + a_0) \cdot 1 = a_k x^k + a_{k-1} x^{k-1} + \cdots + a_1 x + a_0.$$

Now, we check associativity. Let $A, B, D \in \mathbb{Z}[x]$, where $A = a_k x^k + a_{k-1} x^{k-1} + \cdots + a_1 x + a_0$, $B = b_k x^k + b_{k-1} x^{k-1} + \cdots + b_1 x + b_0$, and $D = d_k x^k + d_{k-1} x^{k-1} + \cdots + d_1 x + d_0$.

Then $AB = C$, where $C = c_k x^k + c_{k-1} x^{k-1} + \cdots + c_1 x + c_0$ with

$$c_i = a_i b_0 + a_{i-1} b_1 + \cdots + a_1 b_{i-1} + a_0 b_i \text{ for each } i = 0, 1, \ldots, k.$$

So, $(AB)D = E$, where $E = e_k x^k + e_{k-1} x^{k-1} + \cdots + e_1 x + e_0$ such that for each $i = 0, 1, \ldots, k$,

$$e_i = c_i d_0 + c_{i-1} d_1 + \cdots + c_1 d_{i-1} + c_0 d_i$$
$$= (a_i b_0 + a_{i-1} b_1 + \cdots + a_0 b_i) d_0 + (a_{i-1} b_0 + a_{i-2} b_1 + \cdots + a_0 b_{i-1}) d_1 + \cdots + (a_0 b_0) d_i.$$

Now, $BD = F$, where $F = f_k x^k + f_{k-1} x^{k-1} + \cdots + f_1 x + f_0$ with

$$f_i = b_i d_0 + b_{i-1} d_1 + \cdots + b_1 d_{i-1} + b_0 d_i \text{ for each } i = 0, 1, \ldots, k.$$

So, $A(BD) = G$, where $G = g_k x^k + g_{k-1} x^{k-1} + \cdots + g_1 x + g_0$ such that for each $i = 0, 1, \ldots, k$,

$$g_i = a_i f_0 + a_{i-1} f_1 + \cdots + a_1 f_{i-1} + a_0 f_i$$
$$= a_i (b_0 d_0) + a_{i-1} (b_1 d_0 + b_0 d_1) + \cdots + a_0 (b_i d_0 + b_{i-1} d_1 + \cdots + b_1 d_{i-1} + b_0 d_i).$$

Since $(AB)D = E$ and $A(BD) = G$, the proof is complete once we verify that $E = G$, or equivalently, for each $i = 0, 1, \ldots, k$,

$$(a_ib_0 + a_{i-1}b_1 + \cdots + a_0b_i)d_0 + (a_{i-1}b_0 + a_{i-2}b_1 + \cdots + a_0b_{i-1})d_1 + \cdots + (a_0b_0)d_i$$
$$= a_i(b_0d_0) + a_{i-1}(b_1d_0 + b_0d_1) + \cdots + a_0(b_id_0 + b_{i-1}d_1 + \cdots + b_1d_{i-1} + b_0d_i).$$

This equation can easily be verified by using the distributive property and rearranging the terms.

The proofs of left and right distributivity are similar (but not as tedious).

$(\mathbb{Z}_n[x], +, \cdot)$ is not closed under multiplication. For example, $x^n \in \mathbb{Z}_n[x]$, but $x^{2n} = x^n x^n \notin \mathbb{Z}_n[x]$. So, $(\mathbb{Z}_n[x], +, \cdot)$ is not a ring.

Each step in the proof that $(\mathbb{Z}[x], +, \cdot)$ used only the ring axioms and not any special properties of $\mathbb{Z}$. Therefore, if R is any ring, then $(R[x], +, \cdot)$ is also a ring. $\square$

Note: Since $\mathbb{Q}$, $\mathbb{R}$, and $\mathbb{C}$ are fields (and therefore, rings), $\mathbb{Q}[x]$, $\mathbb{R}[x]$, and $\mathbb{C}[x]$ are rings.

16. Let N be a normal subgroup of the group G, and define $f: G \to G/N$ by $f(g) = gN$. Prove that f is a surjective group homomorphism with kernel N. Conversely, prove that if $f: G \to H$ is a group homomorphism, then $G/\ker(f) \cong f[G]$.

Proof: Define $f: G \to G/N$ by $f(g) = gN$.

f is surjective because if $gN \in G/N$, then $g \in G$ and $f(g) = gN$.

Now, $f(gh) = (gh)N = (gN)(hN) = f(g)f(h)$. So, f is a homomorphism.

If $g \in N$, then by Problem 12, $gN = eN$ because $ge^{-1} = ge = g \in N$. Also, $eN = N$ because for all $x \in N$, $x = ex$. So, $f(g) = gN = eN = N$. Since N is the identity element of G/N. $g \in \ker(f)$. Since $g \in N$ was arbitrary, $N \subseteq \ker(f)$. If $g \in \ker(f)$, then $f(g) = N$. Since $f(g) = gN$, we have $gN = N$. Since $N = eN$, by Problem 12, we have $g = ge = ge^{-1} \in N$. Since $g \in \ker(f)$ was arbitrary, we have $\ker(f) \subseteq N$. Therefore, $\ker(f) = N$.

Next, let $f: G \to H$ be a group homomorphism. Define $F: G/\ker(f) \to f[G]$ by $F(x \ker(f)) = f(x)$.

f is surjective because if $y \in f[G]$, then there is $x \in G$ with $y = f(x)$. So, $F(x \ker(f)) = f(x) = y$.

If $F(x \ker(f)) = F(y \ker(f))$, then $f(x) = f(y)$. So, $f(xy^{-1}) = f(x)f(y^{-1}) = f(x)(f(y))^{-1} = e_H$. So, $xy^{-1} \in \ker(f)$. By Problem 12, $x \ker(f) = y \ker(f)$. Since $x \ker(f), y \ker(f) \in G/\ker(f)$ were arbitrary, F is injective.

Finally, $F(x \ker(f) \cdot y \ker(f)) = F(xy \ker(f)) = f(xy) = f(x)f(y) = F(x \ker(f)) \cdot F(y \ker(f))$. So, F is a homomorphism. $\square$

17. Let I be an ideal of a ring R, and define $f: R \to R/I$ by $f(x) = x + I$. Prove that f is a surjective ring homomorphism with kernel I. Conversely, prove that if $f: R \to S$ is a ring homomorphism, then $R/\ker(f) \cong f[R]$.

Proof: Define $f: R \to R/I$ by $f(x) = x + I$.

f is surjective because if $x + I \in R/I$, then $x \in R$ and $f(x) = x + I$. Now, we have
$$f(x + y) = (x + y) + I = (x + I) + (y + I) = f(x) + f(y).$$
$$f(xy) = xy + I = (x + I)(y + I) = f(x)f(y).$$

So, f is a homomorphism.

If $x \in I$, then by Problem 13, $x + I = 0 + I$ because $x - 0 = x \in I$. Also, $0 + I = I$ because for all $x \in I$, $x = 0 + x$. So, $f(x) = x + I = 0 + I = I$. Since I is the identity element of R/I, $x \in \ker(f)$. Since $x \in I$ was arbitrary, $I \subseteq \ker(f)$. If $x \in \ker(f)$, then $f(x) = I$. Since $f(x) = x + I$, we have $x + I = I$. Since $I = 0 + I$, by Problem 13, we have $x = x - 0 \in I$. Since $x \in \ker(f)$ was arbitrary, we have $\ker(f) \subseteq I$. Therefore, $\ker(f) = I$.

Next, let $f: R \to S$ be a ring homomorphism. Define $F: R/\ker(f) \to f[R]$ by $F(x + \ker(f)) = f(x)$.

f is surjective because if $y \in f[R]$, then there is $x \in R$ with $y = f(x)$. So, $F(x + \ker(f)) = f(x) = y$.

If $F(x + \ker(f)) = F(y + \ker(f))$, then $f(x) = f(y)$. So, $f(x - y) = f(x) - f(y) = 0$. Therefore, $x - y \in \ker(f)$. So, by Problem 13, $x + \ker(f) = y + \ker(f)$. So, F is injective.

$$F\big((x + \ker(f)) + (y + \ker(f))\big) = F\big((x + y) + \ker(f)\big) = f(x + y)$$
$$= f(x) + f(y) = F(x + \ker(f)) + F(y + \ker(f)).$$
$$F\big((x + \ker(f))(y + \ker(f))\big) = F(xy + \ker(f)) = f(xy)$$
$$= f(x) \cdot f(y) = F(x + \ker(f)) \cdot F(y + \ker(f)).$$

Also, $F(1_R + \ker(f)) = f(1_R) = 1_S$. So, F is a homomorphism. $\square$

18. Prove that $(^{\mathbb{R}}\mathbb{R}, +, \cdot)$ is a ring, where addition and multiplication are defined pointwise. Then prove that for each $x \in \mathbb{R}$, $I_x = \{f \in {}^{\mathbb{R}}\mathbb{R} \mid f(x) = 0\}$ is an ideal of $^{\mathbb{R}}\mathbb{R}$ and the only ideal of $^{\mathbb{R}}\mathbb{R}$ containing I_x and not equal to I_x is $^{\mathbb{R}}\mathbb{R}$.

Proof: For $f, g \in {}^{\mathbb{R}}\mathbb{R}$, we define $f + g$ and fg to be the functions in $^{\mathbb{R}}\mathbb{R}$ such that for all $x \in \mathbb{R}$,
$$(f + g)(x) = f(x) + g(x) \text{ and } (fg)(x) = f(x) \cdot g(x).$$

Then $f + g$ and fg are in $^{\mathbb{R}}\mathbb{R}$, so that $^{\mathbb{R}}\mathbb{R}$ is closed under addition and multiplication. To see that addition and multiplication are associative in $^{\mathbb{R}}\mathbb{R}$, we have for each $f, g \in {}^{\mathbb{R}}\mathbb{R}$ and each $x \in \mathbb{R}$,

$$(f + (g + h))(x) = f(x) + (g + h)(x) = f(x) + \big(g(x) + h(x)\big) = \big(f(x) + g(x)\big) + h(x)$$
$$= (f + g)(x) + h(x) = \big((f + g) + h\big)(x).$$

$$(f(gh))(x) = f(x) \cdot (gh)(x) = f(x) \cdot \big(g(x) \cdot h(x)\big) = \big(f(x) \cdot g(x)\big) \cdot h(x)$$
$$= (fg)(x) \cdot h(x) = \big((fg)h\big)(x).$$

Let $0 \in {}^{\mathbb{R}}\mathbb{R}$ be defined by $0(x) = 0$ for all $x \in \mathbb{R}$ and let $1 \in {}^{\mathbb{R}}\mathbb{R}$ be defined by $1(x) = 1$ for all $x \in \mathbb{R}$. Then for all $x \in \mathbb{R}$, we have

$$(0+f)(x) = 0(x) + f(x) = 0 + f(x) = f(x) \qquad (f+0)(x) = f(x) + 0(x) = f(x) + 0 = f(x)$$
$$(1f)(x) = 1(x) \cdot f(x) = 1f(x) = f(x) \qquad (f \cdot 1)(x) = f(x) \cdot 1(x) = f(x) \cdot 1 = f(x).$$

So, 0 is an additive identity in $^{\mathbb{R}}\mathbb{R}$ and 1 is a multiplicative identity in $^{\mathbb{R}}\mathbb{R}$.

The additive inverse of $f \in {}^{\mathbb{R}}\mathbb{R}$ is $-f$ defined by $(-f)(x) = -f(x)$. Indeed, for each $x \in \mathbb{R}$, we have
$$(f + (-f))(x) = f(x) + (-f)(x) = f(x) + (-f(x)) = 0 = 0(x).$$

To see that addition is commutative in $^{\mathbb{R}}\mathbb{R}$, we have for each $f, g \in {}^{\mathbb{R}}\mathbb{R}$ and each $x \in \mathbb{R}$,
$$(f + g)(x) = f(x) + g(x) = g(x) + f(x) = (g + f)(x).$$

It follows that $(^{\mathbb{R}}\mathbb{R}, +, \cdot)$ is a ring.

Now, let $x \in \mathbb{R}$ and let $I_x = \{f \in {}^{\mathbb{R}}\mathbb{R} \mid f(x) = 0\}$. If $f, g \in I_x$, then we have
$$(f + g)(x) = f(x) + g(x) = 0 + 0 = 0.$$

So, I_x is closed under addition.

Since $0(x) = 0$, $0 \in I_x$. So, 0 is an additive identity element for I_x.

If $f \in I_x$, then $(-f)(x) = -f(x) = -0 = 0$, so that $-f \in I_x$. Therefore, I_x has the additive inverse property.

So, $(I_x, +)$ is a subgroup of $(^{\mathbb{R}}\mathbb{R}, +)$.

Let $f \in I_x$ and $g \in {}^{\mathbb{R}}\mathbb{R}$. Then for each $x \in \mathbb{R}$, we have $(fg)(x) = f(x)g(x) = 0 \cdot g(x) = 0 = 0(x)$ and $(gf)(x) = g(x)f(x) = g(x) \cdot 0 = 0 = 0(x)$. Therefore, $fg \in I_x$ and $gf \in I_x$. So, I_x absorbs $^{\mathbb{R}}\mathbb{R}$. It follows that I_x is an ideal of $^{\mathbb{R}}\mathbb{R}$.

Let K be an ideal containing I_x, such that $K \neq I_x$. Then there is $f \in K \setminus I_x$. Let $g \in {}^{\mathbb{R}}\mathbb{R} \setminus K$ be arbitrary. Define $h, k \in {}^{\mathbb{R}}\mathbb{R}$ by $h(z) = \begin{cases} 0 & \text{if } z = x \\ g(z) & \text{if } z \neq x \end{cases}$ and $k(z) = \begin{cases} \frac{g(z)}{f(z)} & \text{if } z = x \\ 0 & \text{if } z \neq x \end{cases}$. Then, we have $g = h + fk$. To see this, observe that if $z \neq x$, then $(h + fk)(z) = h(z) + f(z)k(z) = g(z) + f(z) \cdot 0 = g(z)$ and if $z = x$, then $(h + fk)(z) = (h + fk)(x) = h(x) + f(x)k(x) = 0 + f(x) \cdot \frac{g(x)}{f(x)} = g(x)$.

Now, $h \in I_x$ because $h(x) = 0$. Since $I_x \subseteq K$, $h \in K$. Since $f \in K$ and K is an ideal, $fk \in K$. Since $(K, +)$ is a group, $g = h + fk \in K$. Since $g \in {}^{\mathbb{R}}\mathbb{R} \setminus K$ was arbitrary, it follows that $K = {}^{\mathbb{R}}\mathbb{R}$. □

Note: If I is an ideal of a ring R such that the only ideal containing I not equal to I itself is R, then I is called a **maximal ideal** of R. In the solution above, we proved that for each $x \in \mathbb{R}$, I_x is a maximal ideal of $^{\mathbb{R}}\mathbb{R}$.

Problem Set 12

LEVEL 1

1. Write each of the following positive integers as a product of prime factors in canonical form:
   (i) 9; (ii) 13; (ii) 21; (iv) 30; (v) 44; (vi) 693; (vii) 67,500; (viii) 384,659; (ix) 9,699,690

Solutions:

   (i) $9 = 3^2$

   (ii) $13 = 13$

   (iii) $21 = 3 \cdot 7$

   (iv) $30 = 2 \cdot 3 \cdot 5$

   (v) $44 = 2^2 \cdot 11$

   (vi) $693 = 3^2 \cdot 7 \cdot 11$

   (vii) $67{,}500 = 2^2 \cdot 3^3 \cdot 5^4$

   (viii) $384{,}659 = 11^3 \cdot 17^2$

   (ix) $9{,}699{,}690 = 2 \cdot 3 \cdot 5 \cdot 7 \cdot 11 \cdot 13 \cdot 17 \cdot 19$

2. List all prime numbers less than 100.

Solution: 2, 3, 5, 7, 11, 13, 17, 19, 23, 29, 31, 37, 41, 43, 47, 53, 59, 61, 67, 71, 73, 79, 83, 89, 97

3. Find the gcd and lcm of each of the following sets of numbers: (i) {4, 6}; (ii) {12, 180}; (iii) {2, 3, 5}; (iv) {14, 21, 77}; (v) {720, 2448, 5400}; (vi) $\{2^{17}5^4 11^9 23,\ 2^5 3^2 7^4 11^3 13\}$

Solutions:

   (i) $\gcd(4, 6) = \mathbf{2};\ \text{lcm}(4, 6) = \mathbf{12}$.

   (ii) $12 | 180$. So, $\gcd(12, 180) = \mathbf{12};\ \text{lcm}(12, 180) = \mathbf{180}$.

   (iii) $\gcd(2, 3, 5) = \mathbf{1};\ \text{lcm}(2, 3, 5) = \mathbf{30}$.

   (iv) $\gcd(14, 21, 77) = \mathbf{7};\ \text{lcm}(14, 21, 77) = 2 \cdot 7 \cdot 3 \cdot 11 = \mathbf{462}$.

   (v) $720 = 2^4 \cdot 3^2 \cdot 5,\ 2448 = 2^4 \cdot 3^2 \cdot 17,\ 5400 = 2^3 \cdot 3^3 \cdot 5^2$.

          $\gcd(720, 2448, 5400) = 2^3 \cdot 3^2 = \mathbf{72}$.

          $\text{lcm}(720, 2448, 5400) = 2^4 \cdot 3^3 \cdot 5^2 \cdot 17 = \mathbf{183{,}600}$.

   (vi) $\gcd(2^{17}5^4 11^9 23,\ 2^5 3^2 7^4 11^3 13) = 2^5 \cdot 11^3$.

          $\text{lcm}(2^{17}5^4 11^9 23,\ 2^5 3^2 7^4 11^3 13) = 2^{17} \cdot 3^2 \cdot 5^4 \cdot 7^4 \cdot 11^9 \cdot 13 \cdot 23$.

LEVEL 2

4. Determine if each of the following numbers is prime: (i) 101; (ii) 399; (iii) 1829; (iv) 1933; (v) 8051; (vi) 13,873; (vii) 65,623

Solutions:

(i) $\sqrt{101} < 11$ and 101 is not divisible by 2, 3, 5, and 7. So, 101 is **prime**.

(ii) $399 = 7 \cdot 57$. So, 399 is **not prime**.

(iii) $1829 = 31 \cdot 59$. So, 1829 is **not prime**.

(iv) $\sqrt{1933} < 44$ and 1933 is not divisible by 2, 3, 5, 7, 11, 13, 17, 19, 23, 29, 31, 37, 41, and 43. So, 1933 is **prime**.

(v) $8051 = 83 \cdot 97$. So, 8051 is **not prime**.

(vi) 13,873 is **prime** (check that 13,873 is not divisible by any prime number less than 117).

(vii) $65,623 = 137 \cdot 479$. So, 65,623 is **not prime**.

5. Use the division algorithm to find the quotient and remainder when 723 is divided by 17.

Solution: $17 \cdot 42 = 714$ and $17 \cdot 43 = 731$. Since $714 < 723 < 731$, the quotient is 42. The remainder is then $723 - 17 \cdot 42 = 723 - 714 = 9$. So, $723 = 17 \cdot \mathbf{42} + \mathbf{9}$. The quotient is $k = \mathbf{42}$ and the remainder is $r = \mathbf{9}$.

6. For $n \in \mathbb{Z}^+$, let $M_n = n! + 1$. Determine if M_n is prime for $n = 1, 2, 3, 4, 5, 6,$ and 7.

Solution: $M_1 = 1! + 1 = 1 + 1 = 2$. So, M_1 is prime. $M_2 = 2! + 1 = 2 + 1 = 3$. So, M_2 is prime. $M_3 = 3! + 1 = 6 + 1 = 7$. So, M_3 is prime. $M_4 = 4! + 1 = 24 + 1 = 25$. Since $5|25$, M_4 is **not** prime. $M_5 = 5! + 1 = 120 + 1 = 121$. Since $11|121$, M_5 is **not** prime. $M_6 = 6! + 1 = 721$. Since $7|721$, M_6 is **not** prime. $M_7 = 7! + 1 = 5040 + 1 = 5041$. Since $71|5041$, M_7 is **not** prime.

LEVEL 3

7. Use the Euclidean Algorithm to find $\gcd(825, 2205)$. Then express $\gcd(825, 2205)$ as a linear combination of 825 and 2205.

Solution:

$$2205 = 825 \cdot 2 + 555$$
$$825 = 555 \cdot 1 + 270$$
$$555 = 270 \cdot 2 + 15$$
$$270 = 15 \cdot 18 + 0$$

So, $\gcd(825, 2205) = \mathbf{15}$.

Now we have,

$$15 = 555 - 270 \cdot 2 = 555 - (825 - 555 \cdot 1) \cdot 2 = 3 \cdot 555 - 2 \cdot 825$$
$$= 3(2205 - 825 \cdot 2) - 2 \cdot 825 = -\mathbf{8 \cdot 825} + \mathbf{3 \cdot 2205}.$$

8. Prove that if $k \in \mathbb{Z}$ with $k > 1$, then $k^3 + 1$ is not prime.

Proof: $k^3 + 1 = (k+1)(k^2 - k + 1)$. Since $k > 1$, $k + 1 > 2$ and $k^2 - k + 1 > 1 - 1 + 1 = 1$. Since neither factor is 1, $k^3 + 1$ is **not** prime. □

9. Prove that $\gcd(a, b) \mid \text{lcm}(a, b)$.

Proof: Since $\gcd(a, b)$ is a divisor of a, $\gcd(a, b) \mid a$. Since $\text{lcm}(a, b)$ is a multiple of a, $a \mid \text{lcm}(a, b)$. By problem 3 in Problem Set 10, $\mid$ is a transitive relation. It follows that $\gcd(a, b) \mid \text{lcm}(a, b)$. □

10. Let $a, b, c \in \mathbb{Z}$. Prove that $\gcd(a, b) = \gcd(a + bc, b)$.

Proof: Let d be a divisor of a and b. Then there are integers j and k such that $a = dj$ and $b = dk$. It follows that $a + bc = dj + (dk)c = d(j + kc)$, so that d is a divisor of $a + bc$.

Now, let e be a divisor of $a + bc$ and b. Then there are integers j and k such that $a + bc = ej$ and $b = ek$. It follows that $a = (a + bc) - bc = ej - (ek)c = e(j - kc)$, so that e is a divisor of a.

So, a and b have the same common divisors as $a + bc$ and b. Therefore $\gcd(ab) = \gcd(a + bc, b)$. □

11. Let $a, b, k, r \in \mathbb{Z}$ with $a = bk + r$. Prove that $\gcd(a, b) = \gcd(r, b)$.

Proof: $\gcd(r, b) = \gcd(a - bk, b) = \gcd(a + b(-k), b) = \gcd(a, b)$ by problem 10. □

LEVEL 4

12. Prove the Euclidean Algorithm: Let $a, b \in \mathbb{Z}^+$ with $a \geq b$. Let $r_0 = a, r_1 = b$. Apply the division algorithm to r_0 and r_1 to find $k_1, r_2 \in \mathbb{Z}^+$ such that $r_0 = r_1 k_1 + r_2$, where $0 \leq r_2 < r_1$. Iterate this process to get $r_j = r_{j+1} k_{j+1} + r_{j+2}$, where $0 \leq r_{j+2} < r_{j+1}$ for $j = 0, 1, \ldots, n - 1$ so that $r_{n+1} = 0$. Then $\gcd(a, b) = r_n$.

Proof: We use the Division Algorithm to get the following sequence of equations:

$$r_0 = r_1 k_1 + r_2, \qquad 0 \leq r_2 < r_1$$
$$r_1 = r_2 k_2 + r_3, \qquad 0 \leq r_3 < r_2$$
$$\cdots$$
$$r_{n-2} = r_{n-1} k_{n-1} + r_n, \qquad 0 \leq r_n < r_{n-1}$$
$$r_{n-1} = r_n k_n + 0$$

Since each remainder r_i is strictly less than the remainder r_{i-1}, after finitely many iterations of the Division Algorithm, we get a remainder of 0.

By problem 10, we have $\gcd(a, b) = \gcd(r_0, r_1) = \gcd(r_1, r_2) = \cdots = \gcd(r_n, 0) = r_n$. □

13. Prove that if $a|c$ and $b|c$, then $\mathrm{lcm}(a,b) \mid c$.

Proof: Suppose that $a|c$ and $b|c$. By the Division Algorithm, there are integers k and r such that $c = \mathrm{lcm}(a,b) \cdot k + r$ with $0 \leq r < \mathrm{lcm}(a,b)$. Therefore, we have $r = c - \mathrm{lcm}(a,b) \cdot k$. Since $a|c$ and $a|\mathrm{lcm}(a,b)$, it follows from Problem 7 in Problem Set 4 that $a|r$. Similarly, $b|r$. So r is a multiple of both a and b. Since $0 \leq r < \mathrm{lcm}(a,b)$, r must be 0. Thus, $c = \mathrm{lcm}(a,b) \cdot k$ and so, $\mathrm{lcm}(a,b)|c$. □

14. Suppose that $a, b \in \mathbb{Z}^+$, $\gcd(a,b) = 1$, and $c|ab$. Prove that there are integers d and e such that $c = de$, $d|a$, and $e|b$.

Proof: Suppose that $a, b \in \mathbb{Z}^+$, $\gcd(a,b) = 1$, and $c|ab$. Let $a = p_0^{a_0} p_1^{a_1} \cdots p_n^{a_n}$ and $b = p_0^{b_0} p_1^{b_1} \cdots p_n^{b_n}$ be complete prime factorizations of a and b, let $d = \gcd(a,c)$, and let $e = \gcd(b,c)$. If $d = p_0^{d_0} p_1^{d_1} \cdots p_n^{d_n}$ and $e = p_0^{e_0} p_1^{e_1} \cdots p_n^{e_n}$ are complete prime factorizations of d and e, respectively, then we have $d_i \leq a_i$ and $e_i \leq b_i$ for each $i = 1, 2, \ldots, n$. It follows that $\gcd(d, e) = 1$, $d|a$, and $e|b$. Every prime power in the prime factorization of c must appear as a factor of d or e. So, $c = de$. □

15. A **prime triple** is a sequence of three prime numbers of the form p, $p+2$, and $p+4$. For example, $3, 5, 7$ is a prime triple. Prove that there are no other prime triples.

Proof: We will show that if $n \in \mathbb{Z}$, then one of $n, n+2, n+4$ is divisible by 3. By the Division Algorithm, there are unique integers k and r such that $n = 3k + r$ and $r = 0, 1,$ or 2. If $r = 0$, then n is divisible by 3. If $r = 1$, then $n + 2 = (3k + 1) + 2 = 3k + 3 = 3(k + 1)$, and so, $n + 2$ is divisible by 3. Finally, if $r = 2$, then $n + 4 = (3k + 2) + 4 = 3k + 6 = 3(k + 2)$, and so, $n + 4$ is divisible by 3.

So, if $p, p+2, p+4$ is a prime triple, then either $p, p+2$, or $p+4$ must be divisible by 3. The only prime number divisible by 3 is 3. So, $p = 3$, $p + 2 = 3$, or $p + 4 = 3$. If $p = 3$, we get the prime triple $3, 5, 7$. If $p + 2 = 3$, then $p = 1$, which is not prime. If $p + 4 = 3$, then $p = -1$, again not prime. □

LEVEL 5

16. If $a, b \in \mathbb{Z}^+$ and $\gcd(a, b) = 1$, find the following: (i) $\gcd(a, a+1)$; (ii) $\gcd(a, a+2)$; (iii) $\gcd(3a + 2, 5a + 3)$; (iv) $\gcd(a + b, a - b)$; (v) $\gcd(a + 2b, 2a + b)$

Solutions:

(i) Since $a + 1 - a = 1$, we see that 1 can be written as a linear combination of a and $a + 1$. Therefore, $\gcd(a, a+1) = 1$.

(ii) Since $a + 2 - a = 2$, we see that 2 can be written as a linear combination of a and $a + 2$. Therefore, $\gcd(a, a+2) \leq 2$. So, $\gcd(a, a+2) = 1$ or 2.

If a is even, then there is $k \in \mathbb{Z}$ such that $a = 2k$. So, $a + 2 = 2k + 2 = 2(k + 1)$. Thus, we see that $2|a$ and $2|a+2$. Therefore, if a is even, $\gcd(a, a+2) = 2$.

If a is odd, then 2 does not divide a. So, $\gcd(a, a+2)$ cannot be 2, and therefore, must be 1.

(iii) Since $5(3a + 2) - 3(5a + 3) = 15a + 10 - 15a - 9 = 1$, we see that 1 can be written as a linear combination of $3a + 2$ and $5a + 3$. So, $\gcd(3a + 2, 5a + 3) = 1$.

(iv) If d divides $a + b$ and $a - b$, then d divides $(a + b) + (a - b) = 2a$ and d divides $(a + b) - (a - b) = 2b$ because by Problem 7 from Problem Set 4, d divides any linear combination of $a + b$ and $a - b$. By Theorem 12.5, $\gcd(2a, 2b)$ can be written as a linear combination of $2a$ and $2b$. So, again by Problem 7 from Problem Set 4, $d \mid \gcd(2a, 2b) = 2$. So, $d = 1$ or $d = 2$.

If a and b are both odd, then both $a + b$ and $a - b$ are even, and so, $\gcd(a + b, a - b) = 2$.

If a and b do not have the same parity (in other words, one is even and the other is odd), then $a + b$ is odd. So, 2 is not a divisor of $a + b$, and therefore, $\gcd(a + b, a - b) = 1$.

a and b cannot both be even because then $\gcd(a, b) \geq 2$.

(v) If d divides $a + 2b$ and $2a + b$, then d divides $2(a + 2b) - (2a + b) = 3b$ and d divides $2(2a + b) - (a + 2b) = 3a$ because once again by Problem 7 from Problem Set 4, d divides any linear combination of $a + 2b$ and $2a + b$. By Theorem 12.5, $\gcd(3a, 3b)$ can be written as a linear combination of $3a$ and $3b$. So, again by Problem 7 from Problem Set 4, $d \mid \gcd(3a, 3b) = 3$. So, $d = 1$ or $d = 3$.

If we let $a = 2$ and $b = 3$, then we have $a + 2b = 2 + 2 \cdot 3 = 2 + 6 = 8$ and $2a + b = 2 \cdot 2 + 3 = 4 + 3 = 7$. So, $\gcd(a + 2b, 2a + b) = \gcd(8, 7) = 1$.

If we let $a = 2$ and $b = 5$, then we have $a + 2b = 2 + 2 \cdot 5 = 2 + 10 = 12$ and $2a + b = 2 \cdot 2 + 5 = 4 + 5 = 9$. So, $\gcd(a + 2b, 2a + b) = \gcd(12, 9) = 3$.

So, both $\gcd(a + 2b, 2a + b) = 1$ and $\gcd(a + 2b, 2a + b) = 3$ can occur.

17. Find the smallest ideal of $\mathbb{Z}$ containing 6 and 15. Find the smallest ideal of $\mathbb{Z}$ containing 2 and 3. In general, find the smallest ideal of $\mathbb{Z}$ containing j and k, where $j, k \in \mathbb{Z}$.

Solutions: Let I be the smallest ideal of $\mathbb{Z}$ containing 6 and 15. Since $3\mathbb{Z}$ is an ideal of $\mathbb{Z}$ containing 6 and 15, $I \subseteq 3\mathbb{Z}$ because I is the smallest such ideal. If $a \in 3\mathbb{Z}$, then there is an integer k such that $a = 3k$. Since $3 = 15 - 6 - 6$, we have $a = 15k - 6k - 6k \in I$ because $(I, +)$ is a subgroup of $(\mathbb{Z}, +)$. So, $3\mathbb{Z} \subseteq I$. Since $I \subseteq 3\mathbb{Z}$ and $3\mathbb{Z} \subseteq I$, we have $I = 3\mathbb{Z}$.

Let I be the smallest ideal of $\mathbb{Z}$ containing 2 and 3. If $a \in \mathbb{Z}$, then $a = 3a - 2a \in I$ because $(I, +)$ is a subgroup of $(\mathbb{Z}, +)$. So, $\mathbb{Z} \subseteq I$, and therefore, $I = \mathbb{Z}$.

Let I be the smallest ideal of $\mathbb{Z}$ containing j and k, and let $d = \gcd(j, k)$. Since $d\mathbb{Z}$ is an ideal of $\mathbb{Z}$ containing j and k, $I \subseteq d\mathbb{Z}$ because I is the smallest such ideal. If $a \in d\mathbb{Z}$, then there is an integer c such that $a = dc$. By Problem 7 from Problem Set 4, we can write d as a linear combination of j and k, say $d = mj + nk$. So, we have $a = (mj + nk)c = mjc + nkc$. $mjc, nkc \in I$ because I absorbs $\mathbb{Z}$. So, $a = mjc + nkc \in I$ because $(I, +)$ is a subgroup of $(\mathbb{Z}, +)$. Therefore, $d\mathbb{Z} \subseteq I$. Since $I \subseteq d\mathbb{Z}$ and $d\mathbb{Z} \subseteq I$, we have $I = d\mathbb{Z}$, where $d = \gcd(j, k)$.

18. Find all subgroups of $(\mathbb{Z}, +)$ and all submonoids of $(\mathbb{Z}, +)$.

Solutions: Let $n \in \mathbb{N}$. Then $(n\mathbb{Z}, +)$ is a subgroup of $(\mathbb{Z}, +)$. To see this, first note that $0 \in n\mathbb{Z}$ because $0 = n \cdot 0$. Now, let $a, b \in n\mathbb{Z}$. Then there are integers j and k such that $a = nj$ and $b = nk$. It follows that $a - b = nj - nk = n(j - k)$. Since $\mathbb{Z}$ is closed under subtraction, $j - k \in \mathbb{Z}$. So, $a - b \in n\mathbb{Z}$. By problem 4 from Problem Set 11, $(n\mathbb{Z}, +)$ is a subgroup of $(\mathbb{Z}, +)$.

We now show that there are no other subgroups of $(\mathbb{Z}, +)$. Let $(A, +)$ be a subgroup of $(\mathbb{Z}, +)$. If $A \neq \{0\}$, then A must contain a positive integer k, for if $j \in A \setminus \{0\}$, then $-j \in A$ and either j or $-j$ must be positive. By the Well Ordering Principle, A contains a least positive integer n. We will show that $A = n\mathbb{Z}$.

First, if $a \in n\mathbb{Z}$, then there is $b \in \mathbb{Z}$ with $a = nb$. If $b > 0$, then $nb \in A$ because A is closed under addition (this can be proved rigorously using the Principle of Mathematical Induction). If $b < 0$, then $nb = -n(-b) \in \mathbb{Z}$ because $n(-b) \in \mathbb{Z}$ and A is closed under taking inverses. If $b = 0$, then we have $nb = n \cdot 0 = 0$, and $0 \in A$. So, $n\mathbb{Z} \subseteq A$.

Now, let $a \in A$. Since A is closed under taking inverses, we can assume that a is nonnegative. Since $\gcd(n, a)$ can be written as a linear combination of n and a (by Theorem 12.5) and A is a subgroup of $\mathbb{Z}$, it follows that $\gcd(n, a) \in A$. Since $\gcd(n, a)$ is a divisor of n, $\gcd(n, a) \leq n$. Since n is the least positive integer in A, $\gcd(n, a) = 0$ or $\gcd(n, a) = n$. If $\gcd(n, a) = 0$, then $a = 0$. If $\gcd(n, a) = n$, then n is a divisor of a. In either case, $a \in n\mathbb{Z}$. So, $A \subseteq n\mathbb{Z}$.

Since $n\mathbb{Z} \subseteq A$ and $A \subseteq n\mathbb{Z}$, we have $A = n\mathbb{Z}$.

Now, since every subgroup of $(\mathbb{Z}, +)$ is a submonoid of $(\mathbb{Z}, +)$, we have that $(n\mathbb{Z}, +)$ is a submonoid of $(\mathbb{Z}, +)$ for every $n \in \mathbb{N}$. We also have that $(n\mathbb{N}, +)$ is a submonoid of $(\mathbb{Z}, +)$ for each $n \in \mathbb{Z}^+$, where $n\mathbb{N} = \{nk \mid k \in \mathbb{N}\}$. To see this, let $a, b \in n\mathbb{N}$. Then there are natural numbers j and k such that $a = nj$ and $b = nk$. It follows that $a + b = nj + nk = n(j + k)$. Since $\mathbb{N}$ is closed under addition, $j + k \in \mathbb{N}$. So, $a + b \in n\mathbb{N}$. Therefore, $n\mathbb{N}$ is closed under addition. Also, $0 \in n\mathbb{N}$ because $0 = n \cdot 0$. So, $(n\mathbb{N}, +)$ is a submonoid of $(\mathbb{Z}, +)$. Similar reasoning shows that $(-n\mathbb{N}, +)$ is a submonoid of $(\mathbb{Z}, +)$ for each $n \in \mathbb{Z}^+$, where $(-n\mathbb{N}, +) = \{-nk \mid k \in \mathbb{N}\}$.

We now show that there are no other submonoids of $(\mathbb{Z}, +)$. Let $(A, +)$ be a submonoid of $(\mathbb{Z}, +)$. If $A = \{0\}$, then $A = 0\mathbb{Z}$, so assume that $A \neq \{0\}$. If A has only nonnegative integers, then $A = n\mathbb{N}$, where n is the least positive integer in A. The reasoning is similar to what was done above. Similarly, if A has only nonpositive integers, then $A = -n\mathbb{N}$, where $-n\mathbb{N}$ is the least positive integer such that $n \in A$.

Finally, if A has both positive and negative integers, let n be the least element of the set $\{|k| \mid k \in A \wedge k \neq 0\}$. Then $A = n\mathbb{Z}$, again by reasoning similar to what was done above. □

Problem Set 13

LEVEL 1

1. Let $f: \mathbb{R} \to \mathbb{R}$ be defined by $f(x) = 5x - 1$. (i) Prove that $\lim_{x \to 3} f(x) = 14$. (ii) Prove that f is continuous on $\mathbb{R}$.

Proofs:

(i) Let $\epsilon > 0$ and let $\delta = \frac{\epsilon}{5}$. Suppose that $0 < |x - 3| < \delta$. Then we have

$$|(5x - 1) - 14| = |5x - 15| = 5|x - 3| < 5\delta = 5 \cdot \frac{\epsilon}{5} = \epsilon.$$

So, $\lim_{x \to 3} f(x) = 14$. □

(ii) Let $a \in \mathbb{R}$. We will show that f is continuous at a. Let $\epsilon > 0$, let $\delta = \frac{\epsilon}{5}$, and let $|x - a| < \delta$. Then

$$|(5x - 1) - (5a - 1)| = |5x - 5a| = 5|x - a| < 5\delta = 5 \cdot \frac{\epsilon}{5} = \epsilon.$$

So, f is continuous at a. Since $a \in \mathbb{R}$ was arbitrary, f is continuous on $\mathbb{R}$. □

2. Let $r, c \in \mathbb{R}$ and let $f: \mathbb{R} \to \mathbb{R}$ be defined by $f(x) = c$. Prove that $\lim_{x \to r}[f(x)] = c$.

Proof: Let $\epsilon > 0$ and let $\delta = 1$. If $0 < |x - r| < \delta$, then $|f(x) - c| = |c - c| = |0| = 0 < \epsilon$. Therefore, $\lim_{x \to r}[f(x)] = c$. □

3. Let $A \subseteq \mathbb{R}$, let $f: A \to \mathbb{R}$, let $r, k \in \mathbb{R}$, and suppose that $\lim_{x \to r}[f(x)]$ is a finite real number. Prove that $\lim_{x \to r}[kf(x)] = k \lim_{x \to r}[f(x)]$.

Proof: Suppose that $\lim_{x \to r}[f(x)] = L$ and let $\epsilon > 0$. First assume that $k \neq 0$. Since $\lim_{x \to r}[f(x)] = L$, there is $\delta > 0$ such that $0 < |x - r| < \delta$ implies $|f(x) - L| < \frac{\epsilon}{|k|}$. Suppose that $0 < |x - r| < \delta$. Then

$$|kf(x) - kL| = |k||f(x) - L| < |k|\frac{\epsilon}{|k|} = \epsilon.$$

So, $\lim_{x \to r}[kf(x)] = kL = k \lim_{x \to r}[f(x)]$.

If $k = 0$, let $\delta = 1$. If $0 < |x - r| < \delta$, then

$$|kf(x) - kL| = |0f(x) - 0L| = |0| = 0 < \epsilon.$$

So, in this case, we also have $\lim_{x \to r}[kf(x)] = kL = k \lim_{x \to r}[f(x)]$. □

Level 2

4. Let $A \subseteq \mathbb{R}$, let $f: A \to \mathbb{R}$, and let $r \in \mathbb{R}$. Prove that f is continuous at r if and only if $\lim_{x \to r}[f(x)] = f(r)$.

Proof: Let $A \subseteq \mathbb{R}$, let $f: A \to \mathbb{R}$, and let $r \in \mathbb{R}$. First suppose that f is continuous at r. Let $\epsilon > 0$. Then there is $\delta > 0$ such that $|x - r| < \delta$ implies $|f(x) - f(r)| < \epsilon$. Let $x \in \mathbb{R}$ satisfy $0 < |x - r| < \delta$. Then $|x - r| < \delta$. So, $|f(x) - f(r)| < \epsilon$. Since $\epsilon > 0$ was arbitrary, $\lim_{x \to r}[f(x)] = f(r)$.

Now, suppose that $\lim_{x \to r}[f(x)] = f(r)$. Let $\epsilon > 0$. Then there is $\delta > 0$ such that $0 < |x - r| < \delta$ implies $|f(x) - f(r)| < \epsilon$. Let $x \in \mathbb{R}$ satisfy $|x - r| < \delta$. Then $0 < |x - r| < \delta$ or $x = r$. If $0 < |x - r| < \delta$, then $|f(x) - f(r)| < \epsilon$. If $x = r$, then $|f(x) - f(r)| = |f(r) - f(r)| = |0| = 0 < \epsilon$. Since $\epsilon > 0$ was arbitrary, f is continuous at r. $\square$

5. Prove that every polynomial function $p: \mathbb{R} \to \mathbb{R}$ is continuous on $\mathbb{R}$.

Proof: Let $r \in \mathbb{R}$. We first show that for all $n \in \mathbb{N}$ with $n \geq 1$, $\lim_{x \to r}[x^k] = r^k$.

Base case ($n = 1$): Let $\epsilon > 0$ be given and let $\delta = \epsilon$. Then $0 < |x - r| < \delta$ implies $|x - r| < \delta = \epsilon$. Since $\epsilon > 0$ was arbitrary, $\lim_{x \to r}[x] = r = r^1$.

Inductive step: Let $k \in \mathbb{N}$ and assume that $\lim_{x \to r}[x^k] = r^k$. By Theorem 13.4, we have

$$\lim_{x \to r}[x^{k+1}] = \lim_{x \to r}[x^k \cdot x] = \lim_{x \to r}[x^k] \cdot \lim_{x \to r}[x] = r^k \cdot r = r^{k+1}.$$

By the Principle of Mathematical Induction, for all $k \in \mathbb{N}$ with $n \geq 1$, $\lim_{x \to r}[x^k] = r^k$.

Now, let $p: \mathbb{R} \to \mathbb{R}$ be a polynomial, say $p(x) = a_n x^n + a_{n-1} x^{n-1} + \cdots + a_1 x + a_0$. By Problem 2, $\lim_{x \to r}[a_0] = a_0$. By the last paragraph and Problem 3, $\lim_{x \to r}[a_k x^k] = a_k \lim_{x \to r}[x^k] = a_k r^k$. Finally, using Theorem 13.3, we have

$$\lim_{x \to r}[p(x)] = \lim_{x \to r}[a_n x^n + a_{n-1} x^{n-1} + \cdots + a_1 x + a_0]$$
$$= \lim_{x \to r}[a_n x^n] + \lim_{x \to r}[a_{n-1} x^{n-1}] + \cdots + \lim_{x \to r}[a_1 x] + \lim_{x \to r}[a_0]$$
$$= a_n r^n + a_{n-1} r^{n-1} + \cdots + a_1 r + a_0 = p(r).$$

By Problem 4, p is continuous at r. Since $r \in \mathbb{R}$ was arbitrary, p is continuous on $\mathbb{R}$. $\square$

Level 3

6. Let $g: \mathbb{R} \to \mathbb{R}$ be defined by $g(x) = 2x^2 - 3x + 7$. (i) Prove that $\lim_{x \to 1} g(x) = 6$. (ii) Prove that g is continuous on $\mathbb{R}$.

Proofs:

(i) Let $\epsilon > 0$ and let $\delta = \min\left\{1, \frac{\epsilon}{3}\right\}$. Suppose that $0 < |x - 1| < \delta$. Then we have $|x - 1| < 1$, so that $-1 < x - 1 < 1$. Adding 1, we get $0 < x < 2$. Multiplying by 2, we have $0 < 2x < 4$. Subtracting 1 gives us $-1 < 2x - 1 < 3$. So, $-3 < 2x - 1 < 3$, and therefore, $|2x - 1| < 3$. Now, we have

$$|(2x^2 - 3x + 7) - 6| = |2x^2 - 3x + 1| = |2x - 1||x - 1| < 3\delta \leq 3 \cdot \frac{\epsilon}{3} = \epsilon.$$

So, $\lim_{x \to 1} g(x) = 6$. □

(ii) Let $a \in \mathbb{R}$. We will show that f is continuous at a. Let $\epsilon > 0$ and let $\delta = \min\left\{1, \frac{\epsilon}{M}\right\}$, where $M = \max\{|4a - 8|, |4a - 4|\}$. Suppose that $|x - a| < \delta$. Then we have $|x - a| < 1$, so that $-1 < x - a < 1$. Adding $2a - 3$, we get $2a - 4 < x + a - 3 < 2a - 2$. Multiplying by 2 yields $4a - 8 < 2(x + a - 3) < 4a - 4$. Therefore, $-M < 2(x + a - 3) < M$, or equivalently, $|2(x + a - 3)| < M$. Now, we have

$$|(2x^2 - 3x + 7) - (2a^2 - 3a + 7)| = |2(x^2 - a^2) - 3(x - a)| = |x - a||2(x + a - 3)|$$
$$< \delta M \leq \frac{\epsilon}{M} \cdot M = \epsilon.$$

So, g is continuous at a. Since $a \in \mathbb{R}$ was arbitrary, g is continuous on $\mathbb{R}$. □

7. Suppose that $f, g: \mathbb{R} \to \mathbb{R}$, $a \in \mathbb{R}$, f is continuous at a, and g is continuous at $f(a)$. Prove that $g \circ f$ is continuous at a.

Proof: Let $f, g: \mathbb{R} \to \mathbb{R}$, let $a \in \mathbb{R}$, and suppose that f is continuous at a and g is continuous at $f(a)$. Let $\epsilon > 0$. Since g is continuous at $f(a)$, there is $\delta_1 > 0$ such that $|y - f(a)| < \delta_1$ implies $|g(y) - g(f(a))| < \epsilon$. Since f is continuous at a, there is $\delta_2 > 0$ such that $|x - a| < \delta_2$ implies $|f(x) - f(a)| < \delta_1$. Now, suppose that $|x - a| < \delta_2$. Then $|f(x) - f(a)| < \delta_1$. It follows that $|g(f(x)) - g(f(a))| < \epsilon$. Since $\epsilon > 0$ was arbitrary, $g \circ f$ is continuous at a. □

LEVEL 4

8. Let $h: \mathbb{R} \to \mathbb{R}$ be defined by $h(x) = \frac{x^3 - 4}{x^2 + 1}$. Prove that $\lim_{x \to 2} h(x) = \frac{4}{5}$.

Proof: Let $\epsilon > 0$ and let $\delta = \min\left\{1, \frac{2\epsilon}{15}\right\}$. Suppose that $0 < |x - 2| < \delta$. Then we have $|x - 2| < 1$, so that $-1 < x - 2 < 1$. Adding 2, we get $1 < x < 3$. So, $23 < 5x^2 + 6x + 12 < 75$ and therefore, $-75 < 5x^2 + 6x + 12 < 75$. So, $|5x^2 + 6x + 12| < 75$. Also, $2 < x^2 + 1 < 10$. In particular, we have $x^2 + 1 > 2$, and so, $\frac{1}{x^2 + 1} < \frac{1}{2}$. Now, we have

$$\left|\frac{x^3 - 4}{x^2 + 1} - \frac{4}{5}\right| = \left|\frac{5(x^3 - 4)}{5(x^2 + 1)} - \frac{4(x^2 + 1)}{5(x^2 + 1)}\right| = \left|\frac{5x^3 - 4x^2 - 24}{5(x^2 + 1)}\right|$$
$$= \frac{|5x^2 + 6x + 12||x - 2|}{5(x^2 + 1)} < \frac{75\delta}{5 \cdot 2} \leq \frac{75}{10} \cdot \frac{2\epsilon}{15} = \epsilon.$$

So, $\lim_{x \to 2} h(x) = \frac{4}{5}$. □

9. Let $k: (0, \infty) \to \mathbb{R}$ be defined by $k(x) = \sqrt{x}$ (i) Prove that $\lim_{x \to 25} k(x) = 5$. (ii) Prove that f is continuous on $(0, \infty)$. (iii) Is f uniformly continuous on $(0, \infty)$?

Proofs:

(i) Let $\epsilon > 0$ and let $\delta = \min\{1, (5 + \sqrt{24})\epsilon\}$. Suppose that $0 < |x - 25| < \delta$. Then we have $|x - 25| < 1$, so that $-1 < x - 25 < 1$. Adding 25, we get $24 < x < 26$. Taking square roots, we have $\sqrt{24} < \sqrt{x} < \sqrt{26}$. Adding 5 gives us $5 + \sqrt{24} < \sqrt{x} + 5 < 5 + \sqrt{26}$. So, $\frac{1}{\sqrt{x}+5} < \frac{1}{5+\sqrt{24}}$. Now, we have

$$\left|\sqrt{x} - 5\right| = \left|\frac{(\sqrt{x} - 5)(\sqrt{x} + 5)}{\sqrt{x} + 5}\right| = \frac{|x - 25|}{\sqrt{x} + 5} < \frac{\delta}{5 + \sqrt{24}} \leq \frac{1}{5 + \sqrt{24}} \cdot (5 + \sqrt{24})\epsilon = \epsilon.$$

So, $\lim_{x \to 25} k(x) = 5$. □

(ii) Let $a \in (0, \infty)$. We will show that f is continuous at a. Let $\epsilon > 0$, let $\delta = \min\{1, \epsilon\sqrt{a}\}$, and let $x \in (0, \infty)$ satisfy $|x - a| < \delta$. Then we have $|x - a| < 1$, so that $-1 < x - a < 1$. Adding a, we get $a - 1 < x < a + 1$. Since $x \in (0, \infty)$, we have $0 < x < a + 1$. Taking square roots, we have $0 < \sqrt{x} < \sqrt{a + 1}$. Adding $\sqrt{a}$ gives us $\sqrt{a} < \sqrt{x} + \sqrt{a} < \sqrt{a + 1} + \sqrt{a}$. Therefore, $\frac{1}{\sqrt{x}+\sqrt{a}} < \frac{1}{\sqrt{a}}$. Now, we have

$$\left|\sqrt{x} - \sqrt{a}\right| = \left|\frac{(\sqrt{x} - \sqrt{a})(\sqrt{x} + \sqrt{a})}{\sqrt{x} + \sqrt{a}}\right| = \frac{|x - a|}{\sqrt{x} + \sqrt{a}} < \frac{\delta}{\sqrt{a}} \leq \frac{\epsilon\sqrt{a}}{\sqrt{a}} = \epsilon.$$

So, f is continuous at a. Since $a \in (0, \infty)$ was arbitrary, f is continuous on $(0, \infty)$. □

(iii) Let $\epsilon > 0$, let $\delta = \epsilon^2$, and let $x, y \in (0, \infty)$ satisfy $|x - y| < \delta$. Then we have

$$\left|\sqrt{x} - \sqrt{y}\right| = \sqrt{(\sqrt{x} - \sqrt{y})^2} \leq \sqrt{|\sqrt{x} - \sqrt{y}||\sqrt{x} + \sqrt{y}|} = \sqrt{|x - y|} < \sqrt{\delta} = \sqrt{\epsilon^2} = \epsilon.$$

So, f is uniformly continuous on $\mathbb{R}$. □

10. Let $f: \mathbb{R} \to \mathbb{R}$ be defined by $f(x) = x^2$. Prove that f is continuous on $\mathbb{R}$, but not uniformly continuous on $\mathbb{R}$.

Proof: Let $a \in \mathbb{R}$. We will show that f is continuous at a. Let $\epsilon > 0$ and let $\delta = \min\left\{1, \frac{\epsilon}{M}\right\}$, where $M = \max\{|2a - 1|, |2a + 1|\}$. Suppose that $|x - a| < \delta$. Then $|x - a| < 1$, so that $-1 < x - a < 1$. Adding $2a$, we get $2a - 1 < x + a < 2a + 1$. So, $-M < x + a < M$, or equivalently, $|x + a| < M$. Now, we have

$$|x^2 - a^2| = |x - a||x + a| < \delta \cdot M \leq \frac{\epsilon}{M} \cdot M = \epsilon.$$

So, f is continuous at a. Since $a \in \mathbb{R}$ was arbitrary, f is continuous on $\mathbb{R}$.

To see that f is not uniformly continuous on $\mathbb{R}$, let $\epsilon = 1$ and let $\delta > 0$. Let $x = \frac{1}{\delta}$ and $y = \frac{1}{\delta} + \frac{\delta}{2}$. Then we have $|x - y| = \left|\frac{1}{\delta} - \left(\frac{1}{\delta} + \frac{\delta}{2}\right)\right| = \frac{\delta}{2}$, but

$$|f(x) - f(y)| = |x^2 - y^2| = \left|\frac{1}{\delta^2} - \left(\frac{1}{\delta} + \frac{\delta}{2}\right)^2\right| = \left|\frac{1}{\delta^2} - \frac{1}{\delta^2} - 1 - \frac{\delta^2}{4}\right| = 1 + \frac{\delta^2}{4} > 1 = \epsilon.$$

So, f is **not** uniformly continuous on $\mathbb{R}$ (and in fact, not uniformly continuous on $(0, \infty)$ since we only needed positive values of x and y to violate the definition of uniform continuity). $\square$

11. Prove that if $\lim_{x \to r}[f(x)] > 0$, then there is a deleted neighborhood N of r such that $f(x) > 0$ for all $x \in N$.

Proof: Suppose that $\lim_{x \to r}[f(x)] = L$ with $L > 0$. Let $\epsilon = \frac{L}{2}$. There is $\delta > 0$ such that $0 < |x - r| < \delta$ implies $|f(x) - L| < \epsilon$. Consider $N_\delta^\odot(r) = (r - \delta, r) \cup (r, r + \delta)$. Let $x \in N_\delta^\odot(r)$. Then we have $x \in (r - \delta, r) \cup (r, r + \delta)$, so that $0 < |x - r| < \delta$. It follows that $|f(x) - L| < \epsilon = \frac{L}{2}$. So, we have $-\frac{L}{2} < f(x) - L < \frac{L}{2}$, or equivalently, $L - \frac{L}{2} < f(x) < L + \frac{L}{2}$. Since $L - \frac{L}{2} = \frac{L}{2}$ and $L + \frac{L}{2} = \frac{3L}{2}$, we have $\frac{L}{2} < f(x) < \frac{3L}{2}$. In particular, we have $f(x) > \frac{L}{2} > 0$. Since $x \in N_\delta^\odot(r)$ was arbitrary, we have shown that for all $x \in N_\delta^\odot(r)$, $f(x) > 0$. $\square$

12. Let $A \subseteq \mathbb{R}$, let $f: A \to \mathbb{R}$, let $r \in \mathbb{R}$, and suppose that $\lim_{x \to r}[f(x)]$ is a finite real number. Prove that there is $M \in \mathbb{R}$ and an open interval (a, b) containing r such that $|f(x)| \leq M$ for all $x \in (a, b) \setminus \{r\}$.

Proof: Let $A \subseteq \mathbb{R}$, $f: A \to \mathbb{R}$, $r \in \mathbb{R}$, $\lim_{x \to r}[f(x)] = L$, and let $\epsilon = 1$. Then there is $\delta > 0$ such that $0 < |x - r| < \delta$ implies $|f(x) - L| < 1$, or $-1 < f(x) - L < 1$, or $L - 1 < f(x) < L + 1$. Let $a = r - \delta$, $b = r + \delta$, and $M = \max\{|L - 1|, |L + 1|\}$. If $x \in (a, b) \setminus \{r\}$, then $r - \delta < x < r + \delta$ and $x \neq r$. So, $0 < |x - r| < \delta$. Therefore, $L - 1 < f(x) < L + 1$. Since $M \geq |L - 1| \geq 1 - L$, we have $-M \leq L - 1$. Also, $M \geq |L + 1| \geq L + 1$. So, we have $-M < f(x) < M$, or equivalently, $|f(x)| < M$. Since $x \in (a, b) \setminus \{r\}$ was arbitrary, $|f(x)| < M$ for all $x \in (a, b) \setminus \{r\}$. $\square$

13. Let $A \subseteq \mathbb{R}$, let $f, g, h: A \to \mathbb{R}$, let $r \in \mathbb{R}$, let $f(x) \leq g(x) \leq h(x)$ for all $x \in A \setminus \{r\}$, and suppose that $\lim_{x \to r}[f(x)] = \lim_{x \to r}[h(x)] = L$. Prove that $\lim_{x \to r}[g(x)] = L$.

Proof: Let $\epsilon > 0$. Since $\lim_{x \to r}[f(x)] = L$, there is $\delta_1 > 0$ such that $0 < |x - r| < \delta_1$ implies $|f(x) - L| < \epsilon$. Since $\lim_{x \to r}[h(x)] = L$, there is $\delta_2 > 0$ such that $0 < |x - r| < \delta_2$ implies $|h(x) - L| < \epsilon$. Let $\delta = \min\{\delta_1, \delta_2\}$ and let $0 < |x - r| < \delta$. Then $0 < |x - r| < \delta_1$, so that $|f(x) - L| < \epsilon$, or equivalently, $-\epsilon < f(x) - L < \epsilon$, or $L - \epsilon < f(x) < L + \epsilon$. We will need only that $L - \epsilon < f(x)$. Similarly, we have $0 < |x - r| < \delta_2$, so that $|h(x) - L| < \epsilon$, or equivalently, $-\epsilon < h(x) - L < \epsilon$, or $L - \epsilon < h(x) < L + \epsilon$. We will need only that $h(x) < L + \epsilon$. Now, we have $L - \epsilon < f(x) \leq g(x) \leq h(x) < L + \epsilon$. So, $-\epsilon < g(x) - L < \epsilon$, or equivalently, $|g(x) - L| < \epsilon$. Since $\epsilon > 0$ was arbitrary, $\lim_{x \to r}[g(x)] = L$. $\square$

LEVEL 5

14. Let $A \subseteq \mathbb{R}$, let $f, g : A \to \mathbb{R}$ such that $g(x) \neq 0$ for all $x \in A$, let $r \in \mathbb{R}$, and suppose that $\lim_{x \to r}[f(x)]$ and $\lim_{x \to r}[g(x)]$ are both finite real numbers such that $\lim_{x \to r}[g(x)] \neq 0$. Prove that $\lim_{x \to r}\left[\frac{f(x)}{g(x)}\right] = \frac{\lim_{x \to r} f(x)}{\lim_{x \to r} g(x)}$.

Proof: Suppose that $\lim_{x \to r}[f(x)] = L$ and $\lim_{x \to r}[g(x)] = K$, and let $\epsilon > 0$. Since $\lim_{x \to r}[g(x)] = K$, there is $\delta_1 > 0$ such that $0 < |x - r| < \delta_1$ implies $|g(x) - K| < \frac{|K|}{2}$. Now, $|g(x) - K| < \frac{|K|}{2}$ is equivalent to $-\frac{|K|}{2} < g(x) - K < \frac{|K|}{2}$, or by adding K, $K - \frac{|K|}{2} < g(x) < K + \frac{|K|}{2}$. If $K > 0$, we have $\frac{K}{2} < g(x) < \frac{3K}{2}$. If $K < 0$, we have $\frac{3K}{2} < g(x) < \frac{K}{2}$. In both cases, we have $\frac{|K|}{2} < |g(x)| < \frac{3|K|}{2}$. Let $M = \frac{|K|}{2}$. Then $|g(x)| > M$, and so, $\frac{1}{|g(x)|} < \frac{1}{M}$.

Now, since $\lim_{x \to r}[f(x)] = L$, there is $\delta_2 > 0$ such that $0 < |x - r| < \delta_2$ implies $|f(x) - L| < \frac{M|K|\epsilon}{|K|+|L|}$. Since $\lim_{x \to r}[g(x)] = K$, there is $\delta_3 > 0$ such that $0 < |x - r| < \delta_3$ implies $|g(x) - K| < \frac{M|K|\epsilon}{|K|+|L|}$. Let $\delta = \min\{\delta_1, \delta_2, \delta_3\}$ and suppose that $0 < |x - r| < \delta$. Then since $\delta \leq \delta_1$, $\frac{1}{|g(x)|} < \frac{1}{M}$. Since $\delta \leq \delta_2$, $|f(x) - L| < \frac{M|K|\epsilon}{|K|+|L|}$. Since $\delta \leq \delta_3$, $|g(x) - K| < \frac{M|K|\epsilon}{|K|+|L|}$. By the Triangle Inequality (and SACT), we have

$$\left|\frac{f(x)}{g(x)} - \frac{L}{K}\right| = \left|\frac{Kf(x) - Lg(x)}{Kg(x)}\right| = \left|\frac{Kf(x) - KL + KL - Lg(x)}{Kg(x)}\right| = \left|\frac{Kf(x) - KL}{Kg(x)} + \frac{KL - Lg(x)}{Kg(x)}\right|$$

$$\leq \left|\frac{Kf(x) - KL}{Kg(x)}\right| + \left|\frac{KL - Lg(x)}{Kg(x)}\right| = \left|\frac{f(x) - L}{g(x)}\right| + \left|\frac{L}{K}\right|\left|\frac{K - g(x)}{g(x)}\right| = \left|\frac{f(x) - L}{g(x)}\right| + \left|\frac{L}{K}\right|\left|\frac{g(x) - K}{g(x)}\right|$$

$$= \frac{1}{|g(x)|}\left(|f(x) - L| + \left|\frac{L}{K}\right||g(x) - K|\right) < \frac{1}{M}\left(\frac{M|K|\epsilon}{|K|+|L|} + \left|\frac{L}{K}\right|\frac{M|K|\epsilon}{|K|+|L|}\right) = \frac{1}{M} \cdot \frac{M|K|\epsilon}{|K|+|L|}\left(1 + \left|\frac{L}{K}\right|\right)$$

$$= \frac{|K|\epsilon}{|K|+|L|}\left(\frac{|K|+|L|}{|K|}\right) = \epsilon.$$

So, $\lim_{x \to r}\left[\frac{f(x)}{g(x)}\right] = \frac{L}{K} = \frac{\lim_{x \to r}[f(x)]}{\lim_{x \to r}[g(x)]}$. $\square$

15. Give a reasonable equivalent definition for each of the following limits (like what was done in Theorem 13.5). r and L are finite real numbers. (i) $\lim_{x \to r} f(x) = -\infty$; (ii) $\lim_{x \to +\infty} f(x) = L$; (iii) $\lim_{x \to -\infty} f(x) = L$; (iv) $\lim_{x \to +\infty} f(x) = +\infty$; (v) $\lim_{x \to +\infty} f(x) = -\infty$; (vi) $\lim_{x \to -\infty} f(x) = +\infty$; (vii) $\lim_{x \to -\infty} f(x) = -\infty$.

Equivalent definitions:

(i) $\lim_{x \to r} f(x) = -\infty$ if and only if $\forall M > 0 \, \exists \delta > 0 \, (0 < |x - r| < \delta \to f(x) < -M)$.

(ii) $\lim_{x \to +\infty} f(x) = L$ if and only if $\forall \epsilon > 0 \, \exists K > 0 \, (x > K \to |f(x) - L| < \epsilon)$.

(iii) $\lim_{x \to -\infty} f(x) = L$ if and only if $\forall \epsilon > 0 \, \exists K > 0 \, (x < -K \to |f(x) - L| < \epsilon)$.

(iv) $\lim_{x \to +\infty} f(x) = +\infty$ if and only if $\forall M > 0 \, \exists K > 0 \, (x > K \to f(x) > M)$.

(v) $\lim_{x \to +\infty} f(x) = -\infty$ if and only if $\forall M > 0 \, \exists K > 0 \, (x > K \to f(x) < -M)$.

(vi) $\lim_{x \to -\infty} f(x) = +\infty$ if and only if $\forall M > 0 \, \exists K > 0 \, (x < -K \to f(x) > M)$.

(vii) $\lim_{x \to -\infty} f(x) = -\infty$ if and only if $\forall M > 0 \, \exists K > 0 \, (x < -K \to f(x) < -M)$.

16. Let $f(x) = -x^2 + x + 1$. Use the $M - K$ definition of an infinite limit (that you came up with in Problem 15) to prove $\lim_{x \to +\infty} f(x) = -\infty$.

Proof: Let $M > 0$ and let $K = \frac{1}{2} + \sqrt{M + \frac{5}{4}}$. Suppose that $x > K$. Then $x - \frac{1}{2} > \sqrt{M + \frac{5}{4}}$, and so, $\left(x - \frac{1}{2}\right)^2 > M + \frac{5}{4}$. So, $x^2 - x + \frac{1}{4} > M + \frac{5}{4}$. Thus, $x^2 - x - 1 > M$. Therefore, $-x^2 + x + 1 < -M$. That is, $f(x) < -M$. So, $\lim_{x \to +\infty} g(x) = -\infty$. □

17. Give a reasonable definition for each of the following limits (like what was done in Theorem 13.6). r and L are finite real numbers. (i) $\lim_{x \to r^-} f(x) = L$; (ii) $\lim_{x \to r^+} f(x) = +\infty$; (iii) $\lim_{x \to r^+} f(x) = -\infty$; (iv) $\lim_{x \to r^-} f(x) = +\infty$; (v) $\lim_{x \to r^-} f(x) = -\infty$.

Definitions:

(i) $\lim_{x \to r^-} f(x) = L$ if and only if $\forall \epsilon > 0 \, \exists \delta > 0 \, (-\delta < x - r < 0 \to |f(x) - L| < \epsilon)$.

(ii) $\lim_{x \to r^+} f(x) = +\infty$ if and only if $\forall M > 0 \, \exists \delta > 0 \, (0 < x - r < \delta \to f(x) > M)$.

(iii) $\lim_{x \to r^+} f(x) = -\infty$ if and only if $\forall M > 0 \, \exists \delta > 0 \, (0 < x - r < \delta \to f(x) < -M)$.

(iv) $\lim_{x \to r^-} f(x) = +\infty$ if and only if $\forall M > 0 \, \exists \delta > 0 \, (-\delta < x - r < 0 \to f(x) > M)$.

(v) $\lim_{x \to r^-} f(x) = -\infty$ if and only if $\forall M > 0 \, \exists \delta > 0 \, (-\delta < x - r < 0 \to f(x) < -M)$.

18. Use the $M - \delta$ definition of a one-sided limit (that you came up with in Problem 17) to prove that $\lim_{x \to 3^-} \frac{1}{x-3} = -\infty$.

Proof: Let $M > 0$ and let $\delta = \frac{1}{M}$. If $-\delta < x - 3 < 0$, then $-\frac{1}{M} < x - 3 < 0$, and so, we have $\frac{1}{x-3} < -M$. Since $M > 0$ was arbitrary, $\lim_{x \to 3^-} \frac{1}{x-3} = -\infty$. □

19. Let $f(x) = \frac{x+1}{(x-1)^2}$. Prove that (i) $\lim_{x \to +\infty} f(x) = 0$; (ii) $\lim_{x \to 1^+} f(x) = +\infty$.

Proofs:

(i) Let $\epsilon > 0$ and let $K = \max\left\{2, 1 + \frac{3}{\epsilon}\right\}$. Let $x > K$. Then $x - 1 > 1 + \frac{3}{\epsilon} - 1 = \frac{3}{\epsilon}$, and therefore, $\frac{1}{x-1} < \frac{\epsilon}{3}$. Also, since $x > 2$, $(x-1)^2 - (x-1) = (x-1)(x-1-1) = (x-1)(x-2) > 0$ (because $x - 1 > 2 - 1 = 1 > 0$ and $x - 2 > 2 - 2 > 0$). Thus, $(x-1)^2 > x - 1$, and so, $\frac{1}{(x-1)^2} < \frac{1}{x-1} < \frac{\epsilon}{3}$. It follows from the triangle inequality (and SACT) that

$$\left|\frac{x+1}{(x-1)^2} - 0\right| = \left|\frac{x-1+2}{(x-1)^2}\right| = \left|\frac{x-1}{(x-1)^2} + \frac{2}{(x-1)^2}\right| = \left|\frac{1}{x-1} + \frac{2}{(x-1)^2}\right|$$

$$\leq \left|\frac{1}{x-1}\right| + \left|\frac{2}{(x-1)^2}\right| = \frac{1}{x-1} + 2\frac{1}{(x-1)^2} < \frac{\epsilon}{3} + 2 \cdot \frac{\epsilon}{3} = 3 \cdot \frac{\epsilon}{3} = \epsilon.$$

So, $\lim\limits_{x \to +\infty} f(x) = 0$. $\square$

(ii) Let $M > 0$ and let $\delta = \min\left\{1, \frac{3}{M}\right\}$. If $0 < x - 1 < \delta$, then $0 < x - 1 < \frac{3}{M}$, and so, we have $\frac{1}{x-1} > \frac{M}{3}$. Since $0 < x - 1 < 1$, $(x-1)^2 < x - 1$, and so, $\frac{1}{(x-1)^2} > \frac{1}{x-1}$. So, we have

$$\frac{x+1}{(x-1)^2} = \frac{x-1+2}{(x-1)^2} = \frac{x-1}{(x-1)^2} + \frac{2}{(x-1)^2} = \frac{1}{x-1} + \frac{2}{(x-1)^2}$$

$$> \frac{1}{x-1} + \frac{2}{x-1} = \frac{3}{x-1} > 3 \cdot \frac{M}{3} = M.$$

So, $\lim\limits_{x \to 1} f(x) = +\infty$. $\square$

20. Let $f: \mathbb{R} \to \mathbb{R}$ be defined by $f(x) = \begin{cases} 0 & \text{if } x \text{ is rational.} \\ 1 & \text{if } x \text{ is irrational.} \end{cases}$ Prove that for all $r \in \mathbb{R}$, $\lim\limits_{x \to r}[f(x)]$ does not exist.

Proof: Let $r \in \mathbb{R}$, let $\epsilon = \frac{1}{2}$, and let $\delta > 0$. By the Density Theorem (Theorem 5.10 from Lesson 5) and Problem 11 from Problem Set 5, there is a rational number x and an irrational number y such that $r < x, y < r + \delta$. So, we have $0 < |x - r| < \delta$ and $0 < |y - r| < \delta$. We also have $f(x) = 0$ and $f(y) = 1$. Let $L \in \mathbb{R}$. If $\lim\limits_{x \to r}[f(x)] = L$, then $|f(x) - L| < \frac{1}{2}$ and $|f(y) - L| < \frac{1}{2}$. But then we would have $|f(x) - f(y)| = |f(x) - L + L - f(y)| \leq |f(x) - L| + |L - f(y)| < \frac{1}{2} + \frac{1}{2} = 1$. However, $|f(x) - f(y)| = |1 - 0| = 1$. Since $1 < 1$ is false, $\lim\limits_{x \to r}[f(x)]$ does not equal L. Since $L \in \mathbb{R}$ was arbitrary, $\lim\limits_{x \to r}[f(x)]$ does not exist. $\square$

Problem Set 14

LEVEL 1

1. Let $f: A \to B$ and let X be a nonempty collection of subsets of B. Prove the following: (i) For any $V \in X$, $f[f^{-1}[V]] \subseteq V$; (ii) $f^{-1}[\cup X] = \cup\{f^{-1}[V] \mid V \in X\}$.

Proofs:

(i) Let $V \in X$ and let $y \in f[f^{-1}[V]]$. Then there is $x \in f^{-1}[V]$ with $y = f(x)$. Since $x \in f^{-1}[V]$, we have $y = f(x) \in V$. Since $y \in f[f^{-1}[V]]$ was arbitrary, $f[f^{-1}[V]] \subseteq V$. □

(ii) $x \in f^{-1}[\cup X]$ if and only if $f(x) \in \cup X$ if and only if there is $V \in X$ such that $f(x) \in V$ if and only if there is $V \in X$ such that $x \in f^{-1}[V]$ if and only if $x \in \cup\{f^{-1}[V] \mid V \in X\}$. Since x was arbitrary, $f^{-1}[\cup X] = \cup\{f^{-1}[V] \mid V \in X\}$. □

2. Let (S, d) be a metric space. Prove that for all $x, y \in S$, $d(x, y) \geq 0$.

Proof: Let (S, d) be a metric space and let $x, y \in S$. Then
$$2d(x, y) = d(x, y) + d(x, y) = d(x, y) + d(y, x) \geq d(x, x) = 0.$$

So, $d(x, y) \geq 0$, as desired □

LEVEL 2

3. Prove that $\mathcal{B} = \{X \subseteq \mathbb{R} \mid \mathbb{R} \setminus X \text{ is finite}\}$ generates a topology $\mathcal{T}$ on $\mathbb{R}$ that is strictly coarser than the standard topology. $\mathcal{T}$ is called the **cofinite topology** on $\mathbb{R}$.

Proof: Let $x \in \mathbb{R}$. Then $X = \mathbb{R} \setminus \{x+1\} \in \mathcal{B}$ because $\mathbb{R} \setminus X = \mathbb{R} \setminus (\mathbb{R} \setminus \{x+1\}) = \{x+1\}$, which is finite and $x \in \mathbb{R} \setminus \{x+1\}$ because $x \neq x+1$. So, $\mathcal{B}$ covers $\mathbb{R}$. Let $x \in \mathbb{R}$, and let $X, Y \in \mathcal{B}$ with $x \in X \cap Y$. Then $\mathbb{R} \setminus X$ and $\mathbb{R} \setminus Y$ are both finite and $\mathbb{R} \setminus (X \cap Y) = (\mathbb{R} \setminus X) \cup (\mathbb{R} \setminus Y)$ (by De Morgan's Law) is the union of two finite sets, thus finite. It follows that $X \cap Y \in \mathcal{B}$. Therefore, $\mathcal{B}$ has the intersection containment property. Since $\mathcal{B}$ covers $\mathbb{R}$ and has the intersection containment property, $\mathcal{B}$ is a basis for a topology $\mathcal{T}$ on $\mathbb{R}$.

If X is in $\mathcal{B}$, then X is a union of finitely many open intervals and therefore, X is open in the standard topology on $\mathbb{R}$. This shows that $\mathcal{T}$ is coarser than the standard topology. Since the standard topology on $\mathbb{R}$ is T_2 and $\mathcal{T}$ is not (see Example 14.7), $\mathcal{T}$ is strictly coarser than the standard topology. □

Notes: (1) See part 3 of Example 14.6 and part 2 of Example 14.7 for more information on the cofinite topology.

(2) The topology generated by $\mathcal{B}$ is simply $\mathcal{B} \cup \{\emptyset\}$. $\mathbb{R}$ is in the basis $\mathcal{B}$ because $\mathbb{R} \setminus \mathbb{R} = \emptyset$, which of course is finite. If $X \subseteq \mathcal{B}$, then $\mathbb{R} \setminus \cup X = \cap \{\mathbb{R} \setminus A \mid A \in X\}$. This is an intersection of finite sets, which is finite. So, $\cup X \in \mathcal{B}$. Finally, if $Y \subseteq \mathcal{T}$ and Y is finite, then $\mathbb{R} \setminus \cap Y = \cup\{\mathbb{R} \setminus A \mid A \in Y\}$. This is a finite union of finite sets, which is finite. So, $\cap Y \in \mathcal{B}$.

It follows that $\mathcal{T}$, the topology generated by $\mathcal{B}$, consists of only the cofinite sets together with the empty set.

(3) Since $(0,1)$ is open in the standard topology of $\mathbb{R}$ and is **not** cofinite, by Note 2, $(0,1)$ is **not** open in the cofinite topology. This gives another proof that the cofinite topology on $\mathbb{R}$ is strictly coarser than the standard topology on $\mathbb{R}$.

> 4. Let $K = \left\{\frac{1}{n} \mid n \in \mathbb{Z}^+\right\}$, $\mathcal{B} = \{(a,b) \mid a,b \in \mathbb{R} \land a < b\} \cup \{(a,b) \setminus K \mid a,b \in \mathbb{R} \land a < b\}$. Prove that $\mathcal{B}$ is a basis for a topology $\mathcal{T}_K$ on $\mathbb{R}$ that is strictly finer than the standard topology on $\mathbb{R}$.

Proof: We already know that the bounded open intervals alone cover $\mathbb{R}$. So, $\mathcal{B}$ covers $\mathbb{R}$. By Problem 6 from Problem Set 6 (part (ii)), the intersection of two open intervals with nonempty intersection is an open interval. Furthermore, we have $(a,b) \cap [(c,d) \setminus K] = [(a,b) \cap (c,d)] \setminus K$ and we have $[(a,b) \setminus K] \cap [(c,d) \setminus K] = [(a,b) \cap (c,d)] \setminus K$. This shows that $\mathcal{B}$ has the intersection containment property. Therefore, $\mathcal{B}$ is a basis for a topology $\mathcal{T}_K$ on $\mathbb{R}$. We already saw in part 4 of Example 14.8 that $\mathcal{T}_K$ is strictly finer than the standard topology. □

LEVEL 3

> 5. Let $(K,\mathcal{T})$ and $(L,\mathcal{U})$ be topological spaces with $(K,\mathcal{T})$ compact and let $f: K \to L$ be a homeomorphism. Prove that $(L,\mathcal{U})$ is compact.

Proof: Let $\mathcal{C}$ be an open covering of L. Since f is continuous and bijective, $\mathcal{D} = \{f^{-1}[B] \mid B \in \mathcal{C}\}$ is an open covering of K. Since $(K,\mathcal{T})$ is compact, there is a finite subcollection $\mathcal{E} \subseteq \mathcal{D}$ that covers K. Since f^{-1} is continuous and bijective, $\mathcal{H} = \{f[f^{-1}[B]] \mid f^{-1}[B] \in \mathcal{E}\}$ covers L. By part (i) of Problem 1, $f[f^{-1}[B]] \subseteq B$. Since $\mathcal{H}$ covers L, so does $\mathcal{J} = \{B \mid f^{-1}[B] \in \mathcal{E}\}$. Finally, $\mathcal{J} \subseteq \mathcal{C}$ because if $B \in \mathcal{J}$, then $f^{-1}[B] \in \mathcal{E}$. So, $f^{-1}[B] \in \mathcal{D}$. Therefore, $B \in \mathcal{C}$. □

Note: Since f is surjective, we actually have $f[f^{-1}[B]] = B$. To see this, first note that by Problem 1 above, we have $f[f^{-1}[B]] \subseteq B$. For the other inclusion, let $y \in B$. Since f is surjective, there is $x \in f^{-1}[B]$ with $f(x) = y$. Then $y = f(x) \in f[f^{-1}[B]]$. Since $y \in B$ was arbitrary, we have $B \subseteq f[f^{-1}[B]]$. Since $f[f^{-1}[B]] \subseteq B$ and $B \subseteq f[f^{-1}[B]]$, we have $f[f^{-1}[B]] = B$.

> 6. Let S be a nonempty set and let $\mathcal{B}$ be a collection of subsets of S. Prove that the set generated by $\mathcal{B}$, $\{\bigcup X \mid X \subseteq \mathcal{B}\}$, is equal to $\{A \subseteq S \mid \forall x \in A\, \exists B \in \mathcal{B}(x \in B \land B \subseteq A)\}$.

Proof: Let $\mathcal{C} = \{\bigcup X \mid X \subseteq \mathcal{B}\}$ and let $\mathcal{D} = \{A \subseteq S \mid \forall x \in A\, \exists B \in \mathcal{B}(x \in B \land B \subseteq A)\}$. First, let $A \in \mathcal{C}$. Then there is $X \subseteq \mathcal{B}$ such that $A = \bigcup X$. Let $x \in A$. Then there is a $B \in X$ with $x \in B$. Since $X \subseteq \mathcal{B}$, $B \in \mathcal{B}$. Also, since $B \in X$, $B \subseteq \bigcup X = A$. Therefore, $A \in \mathcal{D}$. Since $A \in \mathcal{C}$ was arbitrary, $\mathcal{C} \subseteq \mathcal{D}$.

Now, let $A \in \mathcal{D}$. For each $x \in A$, there is $B_x \in \mathcal{B}$ such that $x \in B_x$ and $B_x \subseteq A$. If $y \in A$, then $y \in B_y$. So, $y \in \bigcup\{B_x \mid x \in A\}$. So, $A \subseteq \bigcup\{B_x \mid x \in A\}$. If $y \in \bigcup\{B_x \mid x \in A\}$, then $y \in B_x$ for some $x \in A$. Since $B_x \subseteq A$, $y \in A$. Therefore, $\bigcup\{B_x \mid x \in A\} \subseteq A$. It follows that $A = \bigcup\{B_x \mid x \in A\}$. In other words, we have $A = \bigcup X$, where $X = \{B_x \mid x \in A\}$. Since $A \in \mathcal{D}$ was arbitrary, $\mathcal{D} \subseteq \mathcal{C}$.

Since $\mathcal{C} \subseteq \mathcal{D}$ and $\mathcal{D} \subseteq \mathcal{C}$, we have $\mathcal{C} = \mathcal{D}$. □

7. Define the functions d_1 and d_2 from $\mathbb{C} \times \mathbb{C}$ to $\mathbb{R}$ by $d_1(z, w) = |\text{Re } z - \text{Re } w| + |\text{Im } z - \text{Im } w|$ and $d_2(z, w) = \max\{|\text{Re } z - \text{Re } w|, |\text{Im } z - \text{Im } w|\}$. Prove that $(\mathbb{C}, d_1)$ and $(\mathbb{C}, d_2)$ are metric spaces such that d_1 and d_2 induce the standard topology on $\mathbb{C}$.

Proof: $d_1(z, w) = 0$ if and only if $|\text{Re } z - \text{Re } w| + |\text{Im } z - \text{Im } w| = 0$ if and only if $|\text{Re } z - \text{Re } w| = 0$ and $|\text{Im } z - \text{Im } w| = 0$ if and only if $\text{Re } z - \text{Re } w = 0$ and $\text{Im } z - \text{Im } w = 0$ if and only if $\text{Re } z = \text{Re } w$ and $\text{Im } z = \text{Im } w$ if and only if $z = w$. So, property 1 holds for d_1. Property 2 follows immediately from the fact that $|x - y| = |y - x|$ for all $x, y \in \mathbb{R}$. Let's verify property 3. Let $z, w, v \in \mathbb{C}$. Then, we have

$$d_1(z, v) = |\text{Re } z - \text{Re } v| + |\text{Im } z - \text{Im } v|$$

$$= |\text{Re } z - \text{Re } w + \text{Re } w - \text{Re } v| + |\text{Im } z - \text{Im } w + \text{Im } w - \text{Im } v| \text{ (by SACT)}$$

$$\leq |\text{Re } z - \text{Re } w| + |\text{Re } w - \text{Re } v| + |\text{Im } z - \text{Im } w| + |\text{Im } w - \text{Im } v| \text{ (by the Triangle Inequality)}$$

$$= |\text{Re } z - \text{Re } w| + |\text{Im } z - \text{Im } w| + |\text{Re } w - \text{Re } v| + |\text{Im } w - \text{Im } v| = d_1(z, w) + d_1(w, v).$$

This shows that $(\mathbb{C}, d_1)$ is a metric space.

$d_2(z, w) = 0$ if and only if $\max\{|\text{Re } z - \text{Re } w|, |\text{Im } z - \text{Im } w|\} = 0$ if and only if $|\text{Re } z - \text{Re } w| = 0$ and $|\text{Im } z - \text{Im } w| = 0$ if and only if $\text{Re } z - \text{Re } w = 0$ and $\text{Im } z - \text{Im } w = 0$ if and only if $\text{Re } z = \text{Re } w$ and $\text{Im } z = \text{Im } w$ if and only if $z = w$. So, property 1 holds for d_2. Property 2 follows immediately from the fact that $|x - y| = |y - x|$ for all $x, y \in \mathbb{R}$. Let's verify property 3. Let $z, w, v \in \mathbb{C}$. Then, we have

$$d_2(z, v) = \max\{|\text{Re } z - \text{Re } v|, |\text{Im } z - \text{Im } v|\}$$

$$= \max\{|\text{Re } z - \text{Re } w + \text{Re } w - \text{Re } v|, |\text{Im } z - \text{Im } w + \text{Im } w - \text{Im } v|\} \text{ (by SACT)}$$

$$\leq \max\{|\text{Re } z - \text{Re } w| + |\text{Re } w - \text{Re } v|, |\text{Im } z - \text{Im } w| + |\text{Im } w - \text{Im } v|\} \text{ (by the Triangle Inequality)}$$

$$\leq \max\{|\text{Re } z - \text{Re } w|, |\text{Im } z - \text{Im } w|\} + \max\{|\text{Re } w - \text{Re } v|, |\text{Im } w - \text{Im } v|\}$$

(In fact, it's not hard to show that for all $a, b, c, d \in \mathbb{R}$, $\max\{a + b, c + d\} \leq \max\{a, c\} + \max\{b, d\}$)

$$= d_2(z, w) + d_2(w, v).$$

This shows that $(\mathbb{C}, d_2)$ is a metric space.

Let $d: \mathbb{C} \times \mathbb{C} \to \mathbb{R}$ be defined by $d(z, w) = |z - w|$. We have already seen in part 1 of Example 14.10 that d induces the standard topology on $\mathbb{C}$. Let's let $\mathcal{T}$ be the standard topology on $\mathbb{C}$.

Now, if $z, w \in \mathbb{C}$, then we have

$$\max\{|\text{Re } z - \text{Re } w|, |\text{Im } z - \text{Im } w|\} \leq \sqrt{(\text{Re } z - \text{Re } w)^2 + (\text{Im } z - \text{Im } w)^2} = |z - w|$$

$$\leq |\text{Re } z - \text{Re } w| + |\text{Im } z - \text{Im } w| \leq 2\max\{|\text{Re } z - \text{Re } w|, |\text{Im } z - \text{Im } w|\}.$$

Therefore, $d_2(z, w) \leq d(z, w) \leq d_1(z, w) \leq 2d_2(z, w)$.

So, if $z \in \mathbb{C}$ and $r \in \mathbb{R}^+$, then $B_{\frac{r}{2}}(z; d_2) \subseteq B_r(z; d_1) \subseteq B_r(z; d) \subseteq B_r(z; d_2)$. For example, to see that $B_{\frac{r}{2}}(z; d_2) \subseteq B_r(z; d_1)$, if $w \in B_{\frac{r}{2}}(z; d_2)$, then $d_2(z, w) < \frac{r}{2}$, so that $2d_2(z, w) < r$. Then since $d_1(z, w) \leq 2d_2(z, w)$, $d_1(z, w) < r$, so that $w \in B_r(z; d_1)$. The other two arguments are similar.

Let U be an element of the topology induced by d_1. For each $z \in U$, let $B_{r_z}(z; d_1) \subseteq U$. Then for each $z \in U$, $B_{\frac{r_z}{2}}(z; d_2) \subseteq U$. Therefore, $\cup \left\{ B_{\frac{r_z}{2}}(z; d_2) \,\middle|\, z \in U \right\} \subseteq U$. Also, if $w \in U$, then $w \in B_{\frac{r_w}{2}}(w; d_2)$, so that $w \in \cup \left\{ B_{\frac{r_z}{2}}(z; d_2) \,\middle|\, z \in U \right\}$. Therefore, $U \subseteq \cup \left\{ B_{\frac{r_z}{2}}(z; d_2) \,\middle|\, z \in U \right\}$. It follows that we have $U = \cup \left\{ B_{\frac{r_z}{2}}(z; d_2) \,\middle|\, z \in U \right\}$. This shows that the topology $\mathcal{T}_2$ induced by d_2 is finer than the topology $\mathcal{T}_1$ induced by d_1. That is $\mathcal{T}_1 \subseteq \mathcal{T}_2$. Similarly, we have $\mathcal{T} \subseteq \mathcal{T}_1$ and $\mathcal{T}_2 \subseteq \mathcal{T}$. These inclusions together show us that $\mathcal{T} = \mathcal{T}_1 = \mathcal{T}_2$. So, d_1 and d_2 induce the standard topology on $\mathbb{C}$. $\square$

8. Let $(S, \mathcal{T})$ be a topological space and let $A \subseteq S$. Prove that $\mathcal{T}_A = \{A \cap X \mid X \in \mathcal{T}\}$ is a topology on A. Then prove that if $\mathcal{B}$ is a basis for $\mathcal{T}$, then $\mathcal{B}_A = \{A \cap B \mid B \in \mathcal{B}\}$ is a basis for $\mathcal{T}_A$. $\mathcal{T}_A$ is called the **subspace topology** on A.

Proof: $\emptyset = A \cap \emptyset$ shows that $\emptyset \in \mathcal{T}_A$. $A = A \cap S$ shows that $A \in \mathcal{T}_A$.

Now, let $\mathcal{K} \subseteq \mathcal{T}_A$. Then $\mathcal{K} = \{A \cap X \mid X \in \mathcal{H}\}$ for some $\mathcal{H} \subseteq \mathcal{T}$. So,
$$\cup \mathcal{K} = \cup \{A \cap X \mid X \in \mathcal{H}\} = A \cap \cup\{X \mid X \in \mathcal{H}\} = A \cap \cup \mathcal{H}.$$

Since $\mathcal{H} \subseteq \mathcal{T}$ and $\mathcal{T}$ is a topology on S, $\cup \mathcal{H} \in \mathcal{T}$. So, $\cup X = A \cap \cup \mathcal{H} \in \mathcal{T}_A$.

Similarly, if $\mathcal{L} \subseteq \mathcal{T}_A$ is finite, then $\mathcal{L} = \{A \cap X \mid X \in \mathcal{G}\}$ for some finite $\mathcal{G} \subseteq \mathcal{T}$. So,
$$\cap \mathcal{L} = \cap \{A \cap X \mid X \in \mathcal{G}\} = A \cap \cap\{X \mid X \in \mathcal{G}\} = A \cap \cap \mathcal{G}.$$

Since $\mathcal{G} \subseteq \mathcal{T}$ is finite and $\mathcal{T}$ is a topology on S, $\cap \mathcal{G} \in \mathcal{T}$. So, $\cap X = A \cap \cap \mathcal{G} \in \mathcal{T}_A$.

It follows that $\mathcal{T}_A$ is a topology on A.

Since $\mathcal{B}_A \subseteq \mathcal{T}_A$, the set generated by $\mathcal{B}_A$ is contained in the set generated by $\mathcal{T}_A$, which is $\mathcal{T}_A$. Now, let $U \in \mathcal{T}_A$. Then there is $X \in \mathcal{T}$ such that $U = A \cap X$. Since $\mathcal{B}$ is a basis for $\mathcal{T}$, there is $\mathcal{H} \subseteq \mathcal{B}$ such that $X = \cup \mathcal{H}$. Therefore, $U = A \cap \cup \mathcal{H} = \cup\{A \cap X \mid X \in \mathcal{H}\}$, a union of elements from $\mathcal{B}_A$. So, $\mathcal{T}_A$ is contained in the set generated by $\mathcal{B}_A$. It follows that $\mathcal{B}_A$ and $\mathcal{T}_A$ generate the same set. Therefore, $\mathcal{B}_A$ is a basis for $\mathcal{T}_A$. $\square$

Notes: (1) Recall the **generalized distributive law** $A \cap \cup X = \cup\{A \cap B \mid B \in X\}$. This law was introduced in Problem 11 in Problem Set 6. We used this law twice in this proof.

(2) We also used the fact that $\cap\{A \cap X \mid X \in \mathcal{G}\} = A \cap \cap\{X \mid X \in \mathcal{G}\}$. This easy to show, but the dedicated reader should write out the proof in detail.

LEVEL 4

9. Let $\mathcal{B}' = \{(a,b) \mid a,b \in \mathbb{Q} \wedge a < b\}$. Prove that $\mathcal{B}'$ is countable and that $\mathcal{B}'$ is a basis for a topology on $\mathbb{R}$. Then show that the topology generated by $\mathcal{B}'$ is the standard topology on $\mathbb{R}$.

Proof: Define $g: \mathcal{B}' \to \mathbb{Q} \times \mathbb{Q}$ by $g((a,b)) = (a,b)$ (the open interval (a,b) is being sent to the ordered pair (a,b)—it is unfortunate that the notation for these two objects is identical). If $g((a,b)) = g((c,d))$, then $(a,b) = (c,d)$ (as ordered pairs). So, $a = c$ and $b = d$. Therefore, $(a,b) = (c,d)$ (as open intervals). This shows that g is injective. So, $\mathcal{B}' \preccurlyeq \mathbb{Q} \times \mathbb{Q} \sim \mathbb{Q}$. Since $\mathbb{Q}$ is countable, so is $\mathcal{B}'$.

If $x \in \mathbb{R}$, then $x - 1 < x < x + 1$. By the Density Theorem, we can choose $a, b \in \mathbb{Q}$ such that $x - 1 < a < x$ and $x < b < x + 1$. Then $x \in (a,b)$ and $(a,b) \in \mathcal{B}'$. So, $\mathcal{B}'$ covers $\mathbb{R}$. Now, let $x \in \mathbb{R}$ and $(a,b), (c,d) \in \mathcal{B}'$ with $x \in (a,b) \cap (c,d)$. By Problem 6 from Problem Set 6 (part (ii)), we have $(a,b) \cap (c,d) = (e,f)$ for some $e, f \in \mathbb{R}$. By the Density Theorem, we can choose $g, h \in \mathbb{Q}$ such that $e < g < x$ and $x < h < f$. Then $x \in (g,h)$ and $(g,h) \subseteq (a,b) \cap (c,d)$. So, $\mathcal{B}'$ has the intersection containment property. It follows that $\mathcal{B}'$ is a basis for a topology on $\mathbb{R}$.

Since every open interval with rational endpoints is open in the standard topology on $\mathbb{R}$, the topology generated by $\mathcal{B}'$ is contained in the standard topology. Let $a, b \in \mathbb{R}$. For each $n \in \mathbb{Z}^+$, by the Density Theorem, we can choose $q_n, r_n \in \mathbb{Q}$ with $a < q_n < a + \frac{1}{n}$ and $b - \frac{1}{n} < r_n < b$. We will now show that $(a,b) = \bigcup\{(q_n, r_n) \mid n \in \mathbb{Z}^+\}$. If $x \in (a,b)$, then there is $n \in \mathbb{Z}^+$ with $x \in \left(a + \frac{1}{n}, b - \frac{1}{n}\right) \subseteq (q_n, r_n)$. So, $x \in \bigcup\{(q_n, r_n) \mid n \in \mathbb{Z}^+\}$. Therefore, $(a,b) \subseteq \bigcup\{(q_n, r_n) \mid n \in \mathbb{Z}^+\}$. If $x \in \bigcup\{(q_n, r_n) \mid n \in \mathbb{Z}^+\}$, then there is $n \in \mathbb{Z}^+$ such that $x \in (q_n, r_n) \subseteq (a,b)$. So, $\bigcup\{(q_n, r_n) \mid n \in \mathbb{Z}^+\} \subseteq (a,b)$. It follows that $(a,b) = \bigcup\{(q_n, r_n) \mid n \in \mathbb{Z}^+\}$. Since (a,b) is a union of sets in $\mathcal{B}'$, the standard topology is contained in the topology generated by $\mathcal{B}'$. So, $\mathcal{B}'$ generates the standard topology on $\mathbb{R}$. □

10. Let $(S, \mathcal{T})$ be a T_2-space and $A \subseteq S$. Prove that $(A, \mathcal{T}_A)$ is a T_2-space (see Problem 8 for the definition of $\mathcal{T}_A$). Determine if the analogous statement is true for T_3-spaces.

Proof: Let $x, y \in A$ with $x \neq y$. Since $(S, \mathcal{T})$ is a T_2-space, there are $U, V \in \mathcal{T}$ with $x \in U$, $y \in V$, and $U \cap V = \emptyset$. Since $x \in A$ and $x \in U$, $x \in A \cap U$. Since $y \in A$ and $y \in V$, $y \in A \cap V$. By the definition of $\mathcal{T}_A$, $A \cap U$ and $A \cap V$ are in $\mathcal{T}_A$. Finally, $(A \cap U) \cap (A \cap V) = A \cap (U \cap V) = A \cap \emptyset = \emptyset$. Since $x, y \in A$ were arbitrary, $(A, \mathcal{T}_A)$ is a T_2-space.

Let $(S, \mathcal{T})$ be a T_3-space and $A \subseteq S$. We will show that $(A, \mathcal{T}_A)$ is a T_3-space.

Since $(S, \mathcal{T})$ is a T_3-space, $(S, \mathcal{T})$ is also a T_2-space. By the first paragraph, $(A, \mathcal{T}_A)$ is a T_2-space. Since every T_2-space is a T_1-space, $(A, \mathcal{T}_A)$ is a T_1-space.

Let $x \in A$ and $B \subseteq A \setminus \{x\}$ with B closed in $\mathcal{T}_A$. By definition, $A \setminus B$ is open in $\mathcal{T}_A$. So, there is $C \in \mathcal{T}$ with $A \setminus B = A \cap C$. Since $C \in \mathcal{T}$, $S \setminus C$ is closed in $(S, \mathcal{T})$. Since $x \in A \setminus B$, $x \in A \cap C$. So, $x \notin S \setminus C$. Since $(S, \mathcal{T})$ is a T_3-space, there are open sets $U, V \in \mathcal{T}$ with $x \in U$, $S \setminus C \subseteq V$, and $U \cap V = \emptyset$. Since $x \in A$ and $x \in U$, $x \in A \cap U$. Let $b \in B$. Then since $B \subseteq A \setminus \{x\}$, $b \in A$. Since $b \notin A \setminus B$, $b \notin A \cap C$. Since $b \in A$, $b \notin C$. So, $b \in S \setminus C$. Since $b \in B$ was arbitrary, $B \subseteq S \setminus C$. Since $S \setminus C \subseteq V$, $B \subseteq V$. Since $B \subseteq A$ and $B \subseteq V$, $B \subseteq A \cap V$. Finally, $(A \cap U) \cap (A \cap V) = A \cap (U \cap V) = A \cap \emptyset = \emptyset$. Since $x \in A$ was arbitrary and $B \subseteq A \setminus \{x\}$ was an arbitrary closed set, $(A, \mathcal{T}_A)$ is a T_3-space. □

Note: Once again, the topology $\mathcal{T}_A$ in this problem is called the **subspace topology** on A.

11. Let $(S_1, \mathcal{T}_1)$ and $(S_2, \mathcal{T}_2)$ be topological spaces. Let $\mathcal{B} = \{U \times V \mid U \in \mathcal{T}_1 \wedge V \in \mathcal{T}_2\}$. Prove that $\mathcal{B}$ is a basis for a topology $\mathcal{T}$ on $S_1 \times S_2$, but in general, $\mathcal{B}$ itself is not a topology on $S_1 \times S_2$. Then prove that if $\mathcal{B}_1$ is a basis for $\mathcal{T}_1$ and $\mathcal{B}_2$ is a basis for $\mathcal{T}_2$, then $\mathcal{C} = \{U \times V \mid U \in \mathcal{B}_1 \wedge V \in \mathcal{B}_2\}$ is a basis for $\mathcal{T}$. The topology $\mathcal{T}$ is called the **product topology** on $S_1 \times S_2$.

Proof: Let $(x, y) \in S_1 \times S_2$. Since $S_1 \in \mathcal{T}_1$ and $S_2 \in \mathcal{T}_2$, $S_1 \times S_2 \in \mathcal{B}$. So, $\mathcal{B}$ covers $S_1 \times S_2$. Now, let $(x, y) \in (U_1 \times V_1) \cap (U_2 \times V_2)$, where $U_1 \times V_1, U_2 \times V_2 \in \mathcal{B}$. Since $U_1, U_2 \in \mathcal{T}_1$, $U_1 \cap U_2 \in \mathcal{T}_1$. Since $V_1, V_2 \in \mathcal{T}_2$, $V_1 \cap V_2 \in \mathcal{T}_2$. Therefore, $(U_1 \cap U_2) \times (V_1 \cap V_2) \in \mathcal{B}$. $(x, y) \in (U_1 \times V_1) \cap (U_2 \times V_2)$ if and only if $(x, y) \in (U_1 \times V_1)$ and $(x, y) \in (U_2 \times V_2)$ if and only if $x \in U_1$, $y \in V_1$, $x \in U_2$, and $y \in V_2$ if and only if $x \in U_1 \cap U_2$ and $y \in V_1 \cap V_2$ if and only if $(x, y) \in (U_1 \cap U_2) \times (V_1 \cap V_2)$. Therefore, we have $(U_1 \times V_1) \cap (U_2 \times V_2) = (U_1 \cap U_2) \times (V_1 \cap V_2)$. So, $(U_1 \times V_1) \cap (U_2 \times V_2) \in \mathcal{B}$. Therefore, $\mathcal{B}$ has the intersection containment property. It follows that $\mathcal{B}$ is a basis for a topology on $S_1 \times S_2$.

Let S_1 and S_2 be sets, each with at least two elements, let $\mathcal{T}_1$ and $\mathcal{T}_2$ be topologies on S_1 and S_2, respectively, and let $\mathcal{T}$ be the topology generated by $\mathcal{B}$. Let a and b be distinct elements in S_1, let c and d be distinct elements in S_2, let $U_1, U_2 \in \mathcal{T}_1$ with $a \in U_1$, $b \notin U_1$, $a \notin U_2$, $b \in U_2$ and let $V_1, V_2 \in \mathcal{T}_2$ with $c \in V_1$, $d \notin V_1$, $c \notin V_2$, $d \in V_2$. Then the set $X = (U_1 \times V_1) \cup (U_2 \times V_2)$ is in $\mathcal{T}$. However, $X \neq U \times V$ for any $U \in \mathcal{T}_1$ and $V \in \mathcal{T}_2$ because $(a, c) \in X$, $(b, d) \in X$, but $(a, d) \notin X$.

Since $\mathcal{B}_1 \subseteq \mathcal{T}_1$ and $\mathcal{B}_2 \subseteq \mathcal{T}_2$, $\mathcal{C} \subseteq \mathcal{B}$. Therefore, the set generated by $\mathcal{C}$ is contained in the set generated by $\mathcal{B}$, which is $\mathcal{T}$.

Now, let $U \times V \in \mathcal{B}$ and let $(x, y) \in U \times V$. Then $x \in U$ and $y \in V$. Since $U \in \mathcal{T}_1$, there is $U_x \in \mathcal{B}_1$ with $U_x \subseteq U$. Similarly, since $V \in \mathcal{T}_2$, there is $V_y \in \mathcal{B}_2$ with $V_y \subseteq V$. Let $W = \bigcup \{U_x \times V_y \mid x \in U \wedge y \in V\}$. We will show that $U \times V = W$. First, let $(x, y) \in U \times V$. Then $x \in U_x$ and $y \in V_y$. So, $(x, y) \in U_x \times V_y$. Therefore, $(x, y) \in W$. Since $(x, y) \in U \times V$ was arbitrary, $U \times V \subseteq W$. For the reverse inclusion, observe that by construction, for each $x \in U$ and $y \in V$, we have $U_x \times V_y \subseteq U \times V$. It follows that $W \subseteq U \times V$. Since $U \times V \subseteq W$ and $W \subseteq U \times V$, we have $U \times V = W$. This shows that every set in $\mathcal{B}$ is a union of sets in $\mathcal{C}$. Therefore, the set generated by $\mathcal{B}$ (which is $\mathcal{T}$) is contained in the set generated by $\mathcal{C}$. □

Note: The second paragraph in the proof above shows that if $(S_1, \mathcal{T}_1)$ and $(S_2, \mathcal{T}_2)$ are T_1-spaces, each with at least two distinct elements, then $\mathcal{B} = \{U \times V \mid U \in \mathcal{T}_1 \wedge V \in \mathcal{T}_2\}$ is not a topology on $S_1 \times S_2$ (but $\mathcal{B}$ does generate the product topology on $S_1 \times S_2$).

LEVEL 5

12. Let $(S_1, \mathcal{T}_1)$ and $(S_2, \mathcal{T}_2)$ be T_2-spaces. Prove that $S_1 \times S_2$ with the product topology (as defined in Problem 11) is also a T_2-space. Determine if the analogous statement is true for T_3-spaces.

Proof: Let $(x, y), (z, w) \in S_1 \times S_2$ with $(x, y) \neq (z, w)$. Then $x \neq z$ or $y \neq w$. Without loss of generality, assume that $x \neq z$. Since $(S_1, \mathcal{T}_1)$ is a T_2-space, there are $U, V \in \mathcal{T}_1$ with $x \in U$, $z \in V$, and $U \cap V = \emptyset$. Then $(x, y) \in U \times S_1$, $(z, w) \in V \times S_2$, and

$$(U \times S_1) \cap (V \times S_2) = (U \cap V) \times (S_1 \cap S_2) = \emptyset \times (S_1 \cap S_2) = \emptyset.$$

Since $(x, y), (z, w) \in S_1 \times S_2$ were arbitrary, $S_1 \times S_2$ with the product topology is a T_2-space.

Let $(S_1, \mathcal{T}_1)$ and $(S_2, \mathcal{T}_2)$ be T_3-spaces. We will show that $S_1 \times S_2$ with the product topology is also a T_3-space.

Let $(x, y) \in S_1 \times S_2$ and $B \subseteq (S_1 \times S_2) \setminus \{(x, y)\}$ with B closed in the product topology. Consider the open set $(S_1 \times S_2) \setminus B$. Since $(x, y) \in (S_1 \times S_2) \setminus B$, there are sets $U \in \mathcal{T}_1$ and $V \in \mathcal{T}_2$ with $x \in U$, $y \in V$, and $U \times V \subseteq (S_1 \times S_2) \setminus B$. Since U and V are open sets in $(S_1, \mathcal{T}_1)$ and $(S_2, \mathcal{T}_2)$, respectively, $S_1 \setminus U$ and $S_2 \setminus V$ are closed sets in $(S_1, \mathcal{T}_1)$ and $(S_2, \mathcal{T}_2)$, respectively. Also, $x \notin S_1 \setminus U$ and $y \notin S_2 \setminus V$. Since $(S_1, \mathcal{T}_1)$ and $(S_2, \mathcal{T}_2)$ are T_3-spaces, there are open sets W_1, Z_1, W_2, and Z_2 with $x \in W_1$, $y \in W_2$, $S_1 \setminus U \subseteq Z_1$, $S_2 \setminus V \subseteq Z_2$, $W_1 \cap Z_1 = \emptyset$, and $W_2 \cap Z_2 = \emptyset$.

Since $x \in W_1$ and $y \in W_2$, $(x, y) \in W_1 \times W_2$.

We now show that $B \subseteq (Z_1 \times S_2) \cup (S_1 \times Z_2)$. To see this, let $(a, b) \in B$. Then $(a, b) \notin U \times V$. So, $a \notin U$ or $b \notin V$. Without loss of generality, assume that $a \notin U$. Then $a \in Z_1$. So, $(a, b) \in Z_1 \times S_2$. It follows that $(a, b) \in (Z_1 \times S_2) \cup (S_1 \times Z_2)$.

Finally, we have

$$(W_1 \times W_2) \cap [(Z_1 \times S_2) \cup (S_1 \times Z_2)] = [(W_1 \times W_2) \cap (Z_1 \times S_2)] \cup [(W_1 \times W_2) \cap ((S_1 \times Z_2))]$$
$$= [(W_1 \cap Z_1) \times (W_2 \cap S_2)] \cup [(W_1 \cap S_1) \times (W_2 \cap Z_2)]$$
$$= [\emptyset \times (W_2 \cap S_2)] \cup [(W_1 \cap S_1) \times \emptyset] = \emptyset \times \emptyset = \emptyset.$$

Since $(x, y) \in S_1 \times S_2$ was arbitrary and $B \subseteq (S_1 \times S_2) \setminus \{(x, y)\}$ was an arbitrary closed set, $S_1 \times S_2$ with the product topology is a T_3-space. □

13. Let $\mathcal{T}_L$ be the set generated by the half open intervals of the form $[a, b)$ with $a, b \in \mathbb{R}$. Show that $\mathcal{T}_L$ is a topology on $\mathbb{R}$ that is strictly finer than the standard topology on $\mathbb{R}$ and incomparable with the topology $\mathcal{T}_K$.

Proof: Let $\mathcal{B} = \{[a,b) \mid a, b \in \mathbb{R}\}$ and let $\mathcal{T}_L$ be the set generated by $\mathcal{B}$. Let $x \in \mathbb{R}$. Then $x \in [x, x+1)$. This shows that $\mathcal{B}$ covers $\mathbb{R}$. If $[a,b), [c,d) \in \mathcal{B}$ with $[a,b) \cap [c,d) \neq \emptyset$, then $[a,b) \cap [c,d) = [e,f)$, where $e = \max\{a,c\}$ and $f = \min\{b,d\}$. To see this, let $x \in [a,b) \cap [c,d)$. Then we have $a \leq x < b$ and $c \leq x < d$. Since $a \leq x$ and $c \leq x$, $e \leq x$. Since $x < b$ and $x < d$, $x < f$. It follows that $x \in [e,f)$. Conversely, if $x \in [e,f)$, then $e \leq x < f$. Since $a \leq e$ and $c \leq e$, $a \leq x$ and $c \leq x$. Since $f \leq b$ and $f \leq d$, $x < b$ and $x < d$. So, $x \in [a,b)$ and $x \in [c,d)$. Thus, $x \in [a,b) \cap [c,d)$. It follows that $\mathcal{B}$ has the intersection containment property. Since $\mathcal{B}$ covers $\mathbb{R}$ and $\mathcal{B}$ has the intersection containment property, $\mathcal{B}$ is a basis for a topology on $\mathbb{R}$.

To see that $\mathcal{T}_L$ is finer than the standard topology on $\mathbb{R}$, note that each basic open set (a,b) in the standard topology is equal to the union $\cup \left\{ \left[a + \frac{1}{n}, b\right) \mid n \in \mathbb{Z}^+ \right\}$. See the solution to Problem 8 from Problem Set 6 for a proof similar to what is needed to prove this result.

To see that $\mathcal{T}_L$ is **strictly** finer than the standard topology, just note that $[0,1)$ cannot be written as a union of bounded open intervals, for 0 would need to be inside one of those open intervals, and it would then follow that there is an $x < 0$ with $x \in [0,1)$.

The set $(-1, 1) \setminus K$ is open in $\mathcal{T}_K$. We show that $(-1,1) \setminus K$ is **not** open in $\mathcal{T}_L$. If $(-1,1) \setminus K$ is the union of sets of the form $[a,b)$, then 0 would need to be inside one of those half-open intervals, let's say that $0 \in [a,b)$. But then there is some $n > 0$ such that $\frac{1}{n} < b$ (use the Archimedean property). Therefore, $\frac{1}{n} \in [a,b)$. This contradicts that $\frac{1}{n} \in K$, showing that $\mathcal{T}_L$ is **not** finer than $\mathcal{T}_K$.

The set $[0,1)$ is open in $\mathcal{T}_L$. We've already seen that $[0,1)$ cannot be written as a union of bounded open intervals. If we throw additional sets of the form $(a,b) \setminus K$ into such a union, then we still run into the same issue with 0. If $0 \in (a,b)$ or $(a,b) \setminus K$, we would get an $x < 0$ with $x \in [0,1)$. □

14. Prove that every metrizable space is T_4.

Proof: Let $(S, \mathcal{T})$ be metrizable and let d be a metric on S that induces $\mathcal{T}$. Let A, B be disjoint closed subsets of S. Let $x \in A$. Since $A \cap B = \emptyset$, $x \notin B$. So, x is in the open set $S \setminus B$. Therefore, there is $r_x \in \mathbb{R}^+$ such that $B_{r_x}(x) \subseteq S \setminus B$. Let $U = \cup \left\{ B_{\frac{r_x}{2}}(x) \mid x \in A \right\}$. Then $U \in \mathcal{T}$, $A \subseteq U$, and $U \cap B = \emptyset$. Similarly, for each $x \in B$, let $V = \cup \left\{ B_{\frac{r_x}{2}}(x) \mid x \in B \right\}$, so that $V \in \mathcal{T}$, $B \subseteq V$, and $V \cap A = \emptyset$.

We now show that $U \cap V = \emptyset$. If $a \in U \cap V$, then there is $x \in A$ and $y \in B$ with $a \in B_{\frac{r_x}{2}}(x) \cap B_{\frac{r_y}{2}}(y)$. Then $d(x,y) \leq d(x,a) + d(a,y) < \frac{r_x}{2} + \frac{r_y}{2}$. Without loss of generality, assume that $r_y \leq r_x$. Then, we have $d(x,y) < \frac{r_x}{2} + \frac{r_x}{2} = r_x$. So, $y \in B_{r_x}(x)$. Since $B_{r_x}(x) \subseteq S \setminus B$, $y \in S \setminus B$. So, $y \notin B$, a contradiction. It follows that $U \cap V = \emptyset$. □

15. Consider the topological space $(\mathbb{R}, \mathcal{T}_L)$. Prove that $\mathbb{R}^2$ with the corresponding product topology (as defined in Problem 11) is a T_3-space, but not a T_4-space.

Proof: We first show that $(\mathbb{R}, \mathcal{T}_L)$ is a T_4-space. Since $\mathcal{T}_L$ is finer than the standard topology on $\mathbb{R}$, and the standard topology is a T_1-space, $\mathcal{T}_L$ is also a T_1-space. Now, let A, B be disjoint closed subsets of $\mathbb{R}$. For each $a \in A$, we have $a \notin B$. So, $a \in \mathbb{R} \setminus B$. Since $\mathbb{R} \setminus B$ is open, there is a basic open set $[c, x_a)$ containing a such that $[c, x_a) \subseteq \mathbb{R} \setminus B$. Since $c \leq a$, we have $[a, x_a) \subseteq [c, x_a)$, and therefore, $[a, x_a) \subseteq \mathbb{R} \setminus B$. So, $[a, x_a) \cap B = \emptyset$. Similarly, for each $b \in B$, we can find x_b so that $[b, x_b) \cap A = \emptyset$. Let $U = \bigcup \{[a, x_a) \mid a \in A\}$ and let $V = \bigcup \{[b, x_b) \mid b \in B\}$. U and V are unions of basic open sets, thus open. Clearly, $A \subseteq U$ and $B \subseteq V$.

We show that $U \cap V = \emptyset$. Suppose toward contradiction that $z \in U \cap V$. Then there is $a \in A$ and $b \in B$ with $z \in [a, x_a)$ and $z \in [b, x_b)$. Without loss of generality, assume that $a < b$. Since $z \in [b, x_b)$, we have $b \leq z$. Since $z \in [a, x_a)$, we have $z < x_a$. So, $a < b \leq z < x_a$. It follows that $b \in [a, x_a)$, contradicting $[a, x_a) \cap B = \emptyset$. This contradiction shows that $U \cap V = \emptyset$. So, $(\mathbb{R}, \mathcal{T}_L)$ is a T_4-space.

Since $(\mathbb{R}, \mathcal{T}_L)$ is a T_4-space, it is also a T_3-space.

Let $\mathcal{T}$ be the product topology on $\mathbb{R}^2$ with respect to the topology $\mathcal{T}_L$. By the proof of Problem 12, $(\mathbb{R}^2, \mathcal{T})$ is a T_3-space.

We will now show that $(\mathbb{R}^2, \mathcal{T})$ is **not** a T_4-space.

Assume toward contradiction that $(\mathbb{R}^2, \mathcal{T})$ is a T_4-space. Let $D = \{(x, -x) \mid x \in \mathbb{R}\}$. D is a closed set in the standard product topology on $\mathbb{R}$ (as are all lines). Since $\mathcal{T}_L$ is finer than the standard topology on $\mathbb{R}$, D is also closed in $(\mathbb{R}^2, \mathcal{T})$.

Furthermore, $\mathcal{T}_D$ is the discrete topology on D. To see this, observe that the point $(x, -x)$ is equal to the intersection of D with the basic open set $[x, x+1) \times [-x, -x+1)$. Therefore, every singleton set $\{(x, -x)\}$ is open in $\mathcal{T}_D$. It follows that all subsets of D are both open and closed in $(D, \mathcal{T}_D)$.

If $A \subseteq D$, since A is closed in $(D, \mathcal{T}_D)$ and D is closed in $(\mathbb{R}^2, \mathcal{T})$, it follows that, A is closed in $(\mathbb{R}^2, \mathcal{T})$ (Why?). So, for any $A \subseteq D$ with $A \neq \emptyset$ and $A \neq D$, both A and $D \setminus A$ are closed in $(\mathbb{R}^2, \mathcal{T})$. Since we are assuming that $(\mathbb{R}^2, \mathcal{T})$ is a T_4-space, we can find disjoint $U_A, V_A \in \mathcal{T}$ with $A \subseteq U_A$ and $D \setminus A \subseteq V_A$.

Define $f: \mathcal{P}(D) \to \mathcal{P}(\mathbb{Q} \times \mathbb{Q})$ by $f(\emptyset) = \emptyset$, $f(D) = \mathbb{Q} \times \mathbb{Q}$, and $f(A) = (\mathbb{Q} \times \mathbb{Q}) \cap U_A$ for $A \neq \emptyset$ and $A \neq D$. We show that f is injective.

Let $A, B \in \mathcal{P}(D)$, both nonempty, both not equal to D or each other. Without loss of generality, assume there is $(x, -x) \in A \setminus B$. Then $(x, -x) \in D \setminus B$. Therefore, $(x, -x) \in U_A \cap V_B$. Since $U_A \cap V_B$ is open and nonempty, by the Density Theorem, there is $(a, -a) \in (U_A \cap V_B) \cap (\mathbb{Q} \times \mathbb{Q})$. Therefore, we have $(a, -a) \in (\mathbb{Q} \times \mathbb{Q}) \cap U_A = f(A)$ and $(a, -a) \notin (\mathbb{Q} \times \mathbb{Q}) \cap U_B = f(B)$. So, $f(A) \neq f(B)$.

Also, note that if $A \in \mathcal{P}(D)$, with $A \neq \emptyset$ and $A \neq D$, then $f(A) = (\mathbb{Q} \times \mathbb{Q}) \cap U_A$ is not empty because $\mathbb{Q} \times \mathbb{Q}$ has nonempty intersection with any open set, and $f(A) = (\mathbb{Q} \times \mathbb{Q}) \cap U_A$ is not $\mathbb{Q} \times \mathbb{Q}$ because $(\mathbb{Q} \times \mathbb{Q}) \cap V_A \neq \emptyset$. It follows that f is injective.

So, $\mathcal{P}(D) \preccurlyeq \mathcal{P}(\mathbb{Q} \times \mathbb{Q}) \sim \mathcal{P}(\mathbb{Q}) \sim \mathcal{P}(\mathbb{N}) \sim \mathbb{R} \sim D$, contradicting Cantor's Theorem. $\square$

16. Let $(S_1, \mathcal{T}_1)$ and $(S_2, \mathcal{T}_2)$ be metrizable spaces. Prove that $S_1 \times S_2$ with the product topology is metrizable. Use this to show that $(\mathbb{R}, \mathcal{T}_L)$ is not metrizable.

Proof: Let d_1 and d_2 be metrics that induce the topologies $\mathcal{T}_1$ and $\mathcal{T}_2$, respectively. Define $d: (S_1 \times S_2) \times (S_1 \times S_2) \to \mathbb{R}$ by $d((a,b),(c,d)) = \max\{d_1(a,c), d_2(b,d)\}$. We first show that d defines a metric on $S_1 \times S_2$. We have $d((a,b),(c,d)) = 0$ if and only if $\max\{d_1(a,c), d_2(b,d)\} = 0$ if and only if $d_1(a,c) = 0$ and $d_2(b,d) = 0$ if and only if $a = c$ and $b = d$ if and only if $(a,b) = (c,d)$. So, property 1 holds. Property 2 is clear. For property 3, Let $(a,b),(c,k),(e,f) \in S_1 \times S_2$, Then

$$d((a,b),(e,f)) = \max\{d_1(a,e), d_2(b,f)\} \le \max\{d_1(a,c) + d_1(c,e), d_2(b,k) + d_2(k,f)\}$$
$$\le \max\{d_1(a,c), d_2(b,k)\} + \max\{d_1(c,e), d_2(k,f)\} = d((a,b),(c,k)) + d((c,k),(e,f)).$$

We now show that d induces the product topology on $S_1 \times S_2$.

Let $B = B_r((x,y); d)$ be an arbitrary open ball in the topology induced by d. We show that $B = B_r(x; d_1) \times B_r(y; d_2)$. If $(a,b) \in B$, then $\max\{d_1(x,a), d_2(y,b)\} = d((x,y),(a,b)) < r$. So, $d_1(x,a) < r$ and $d_2(y,b) < r$. Thus, $a \in B_r(x; d_1)$, $b \in B_r(y; d_2)$. So, $(a,b) \in B_r(x; d_1) \times B_r(y; d_2)$. Therefore, $B \subseteq B_r(x; d_1) \times B_r(y; d_2)$. Now, if $(a,b) \in B_r(x; d_1) \times B_r(y, d_2)$, then $a \in B_r(x; d_1)$ and $b \in B_r(y; d_2)$. So, $d_1(x,a) < r$ and $d_2(y,b) < r$. So, $d((x,y),(a,b)) = \max\{d_1(x,a), d_2(y,b)\} < r$. Therefore, $(a,b) \in B$. So, $B_r(x; d_1) \times B_r(y; d_2) \subseteq B$. Since $B \subseteq B_r(x; d_1) \times B_r(y, d_2)$ and $B_r(x; d_1) \times B_r(y; d_2) \subseteq B$, we have $B \subseteq B_r(x; d_1) \times B_r(y, d_2)$. This shows that B is open in the product topology. Therefore, the product topology is finer than the topology induced by d.

Conversely, a basic open set in the product topology has the form $B_r(x; d_1) \times B_r(y; d_2)$, and we saw that this equal to $B_r((x,y); d)$. It follows that each basic open set in the product topology is open in the topology induced by d. Therefore, the topology induced by d is finer than the product topology. So, d induces the product topology.

Now, assume towards contradiction that $(\mathbb{R}, \mathcal{T}_L)$ is metrizable. It follows that $\mathbb{R}^2$ with the corresponding product topology is metrizable. By Problem 14, $\mathbb{R}^2$ with the product topology is a T_4-space, contradicting Problem 15. So, $(\mathbb{R}, \mathcal{T}_L)$ is **not** metrizable. $\square$

Problem Set 15

LEVEL 1

1. In Problems 11 and 12 below, you will be asked to show that $W\left(\frac{\pi}{3}\right) = \left(\frac{1}{2}, \frac{\sqrt{3}}{2}\right)$ and $W\left(\frac{\pi}{6}\right) = \left(\frac{\sqrt{3}}{2}, \frac{1}{2}\right)$. Use this information to compute the sine, cosine, and tangent of each of the following angles: (i) $\frac{\pi}{6}$; (ii) $\frac{\pi}{3}$; (iii) $\frac{2\pi}{3}$; (iv) $\frac{5\pi}{6}$; (v) $\frac{7\pi}{6}$; (vi) $\frac{4\pi}{3}$; (vii) $\frac{5\pi}{3}$; (viii) $\frac{11\pi}{6}$.

Solutions:

(i) By Problem 12, $W\left(\frac{\pi}{6}\right) = \left(\frac{\sqrt{3}}{2}, \frac{1}{2}\right)$. So, $\cos\frac{\pi}{6} = \frac{\sqrt{3}}{2}$, $\sin\frac{\pi}{6} = \frac{1}{2}$, and $\tan\frac{\pi}{6} = \frac{\sin\frac{\pi}{6}}{\cos\frac{\pi}{6}} = \frac{1}{2} \cdot \frac{2}{\sqrt{3}} = \frac{1}{\sqrt{3}}$.

(ii) By Problem 11, $W\left(\frac{\pi}{3}\right) = \left(\frac{1}{2}, \frac{\sqrt{3}}{2}\right)$. So, $\cos\frac{\pi}{3} = \frac{1}{2}$, $\sin\frac{\pi}{3} = \frac{\sqrt{3}}{2}$, and $\tan\frac{\pi}{3} = \frac{\sin\frac{\pi}{3}}{\cos\frac{\pi}{3}} = \frac{\sqrt{3}}{2} \cdot \frac{2}{1} = \sqrt{3}$.

(iii) Since $\frac{2\pi}{3} = \pi - \frac{\pi}{3}$, by the symmetry of the unit circle, $W\left(\frac{2\pi}{3}\right) = \left(-\frac{1}{2}, \frac{\sqrt{3}}{2}\right)$. It follows that $\cos\frac{2\pi}{3} = -\frac{1}{2}$, $\sin\frac{2\pi}{3} = \frac{\sqrt{3}}{2}$, and $\tan\frac{2\pi}{3} = \frac{\sin\frac{2\pi}{3}}{\cos\frac{2\pi}{3}} = \frac{\sqrt{3}}{2}\left(\frac{-2}{1}\right) = -\sqrt{3}$.

(iv) Since $\frac{5\pi}{6} = \pi - \frac{\pi}{6}$, by the symmetry of the unit circle, $W\left(\frac{5\pi}{6}\right) = \left(-\frac{\sqrt{3}}{2}, \frac{1}{2}\right)$. It follows that $\cos\frac{5\pi}{6} = -\frac{\sqrt{3}}{2}$, $\sin\frac{5\pi}{6} = \frac{1}{2}$, and $\tan\frac{5\pi}{6} = \frac{\sin\frac{5\pi}{6}}{\cos\frac{5\pi}{6}} = \frac{1}{2}\left(-\frac{2}{\sqrt{3}}\right) = -\frac{1}{\sqrt{3}}$.

(v) Since $\frac{7\pi}{6} = \pi + \frac{\pi}{6}$, by the symmetry of the unit circle, $W\left(\frac{7\pi}{6}\right) = \left(-\frac{\sqrt{3}}{2}, -\frac{1}{2}\right)$. It follows that $\cos\frac{7\pi}{6} = -\frac{\sqrt{3}}{2}$, $\sin\frac{7\pi}{6} = -\frac{1}{2}$, and $\tan\frac{7\pi}{6} = \frac{\sin\frac{7\pi}{6}}{\cos\frac{7\pi}{6}} = -\frac{1}{2}\left(-\frac{2}{\sqrt{3}}\right) = \frac{1}{\sqrt{3}}$.

(vi) Since $\frac{4\pi}{3} = \pi + \frac{\pi}{3}$, by the symmetry of the unit circle, $W\left(\frac{4\pi}{3}\right) = \left(-\frac{1}{2}, -\frac{\sqrt{3}}{2}\right)$. It follows that $\cos\frac{4\pi}{3} = -\frac{1}{2}$, $\sin\frac{4\pi}{3} = -\frac{\sqrt{3}}{2}$, and $\tan\frac{4\pi}{3} = \frac{\sin\frac{4\pi}{3}}{\cos\frac{4\pi}{3}} = -\frac{\sqrt{3}}{2}\left(\frac{-2}{1}\right) = \sqrt{3}$.

(vii) Since $\frac{5\pi}{3} = 2\pi - \frac{\pi}{3}$, by the symmetry of the unit circle, $W\left(\frac{5\pi}{3}\right) = \left(\frac{1}{2}, -\frac{\sqrt{3}}{2}\right)$. It follows that $\cos\frac{5\pi}{3} = \frac{1}{2}$, $\sin\frac{5\pi}{3} = -\frac{\sqrt{3}}{2}$, and $\tan\frac{5\pi}{3} = \frac{\sin\frac{5\pi}{3}}{\cos\frac{5\pi}{3}} = -\frac{\sqrt{3}}{2} \cdot \frac{2}{1} = -\sqrt{3}$.

(viii) Since $\frac{11\pi}{6} = 2\pi - \frac{\pi}{6}$, by the symmetry of the unit circle, $W\left(\frac{11\pi}{6}\right) = \left(\frac{\sqrt{3}}{2}, -\frac{1}{2}\right)$. It follows that $\cos\frac{11\pi}{6} = \frac{\sqrt{3}}{2}$, $\sin\frac{11\pi}{6} = -\frac{1}{2}$, and $\tan\frac{11\pi}{6} = \frac{\sin\frac{11\pi}{6}}{\cos\frac{11\pi}{6}} = -\frac{1}{2} \cdot \frac{2}{\sqrt{3}} = -\frac{1}{\sqrt{3}}$.

2. Use the sum identities (Theorem 15.1) to compute the cosine, sine, and tangent of each of the following angles: (i) $\frac{5\pi}{12}$; (ii) $\frac{\pi}{12}$; (iii) $\frac{11\pi}{12}$; (iv) $\frac{19\pi}{12}$.

Solutions:

(i) $\cos\frac{5\pi}{12} = \cos\left(\frac{\pi}{4} + \frac{\pi}{6}\right) = \cos\frac{\pi}{4}\cos\frac{\pi}{6} - \sin\frac{\pi}{4}\sin\frac{\pi}{6} = \frac{\sqrt{2}}{2}\cdot\frac{\sqrt{3}}{2} - \frac{\sqrt{2}}{2}\cdot\frac{1}{2} = \frac{\sqrt{6}-\sqrt{2}}{4}.$

$\sin\frac{5\pi}{12} = \sin\left(\frac{\pi}{4} + \frac{\pi}{6}\right) = \sin\frac{\pi}{4}\cos\frac{\pi}{6} + \cos\frac{\pi}{4}\sin\frac{\pi}{6} = \frac{\sqrt{2}}{2}\cdot\frac{\sqrt{3}}{2} + \frac{\sqrt{2}}{2}\cdot\frac{1}{2} = \frac{\sqrt{6}+\sqrt{2}}{4}.$

$\tan\frac{5\pi}{12} = \frac{\sin\frac{5\pi}{12}}{\cos\frac{5\pi}{12}} = \frac{\sqrt{6}+\sqrt{2}}{4} \cdot \frac{4}{\sqrt{6}-\sqrt{2}} = \frac{\sqrt{6}+\sqrt{2}}{\sqrt{6}-\sqrt{2}}.$

(ii) $\cos\frac{\pi}{12} = \cos\left(\frac{\pi}{4} - \frac{\pi}{6}\right) = \cos\frac{\pi}{4}\cos\frac{\pi}{6} + \sin\frac{\pi}{4}\sin\frac{\pi}{6} = \frac{\sqrt{2}}{2}\cdot\frac{\sqrt{3}}{2} + \frac{\sqrt{2}}{2}\cdot\frac{1}{2} = \frac{\sqrt{6}+\sqrt{2}}{4}.$

$\sin\frac{\pi}{12} = \sin\left(\frac{\pi}{4} - \frac{\pi}{6}\right) = \sin\frac{\pi}{4}\cos\frac{\pi}{6} - \cos\frac{\pi}{4}\sin\frac{\pi}{6} = \frac{\sqrt{2}}{2}\cdot\frac{\sqrt{3}}{2} - \frac{\sqrt{2}}{2}\cdot\frac{1}{2} = \frac{\sqrt{6}-\sqrt{2}}{4}.$

$\tan\frac{\pi}{12} = \frac{\sin\frac{\pi}{12}}{\cos\frac{\pi}{12}} = \frac{\sqrt{6}-\sqrt{2}}{4} \cdot \frac{4}{\sqrt{6}+\sqrt{2}} = \frac{\sqrt{6}-\sqrt{2}}{\sqrt{6}+\sqrt{2}}.$

(iii) $\cos\frac{11\pi}{12} = \cos\left(\frac{\pi}{4} + \frac{2\pi}{3}\right) = \cos\frac{\pi}{4}\cos\frac{2\pi}{3} - \sin\frac{\pi}{4}\sin\frac{2\pi}{3} = \frac{\sqrt{2}}{2}\cdot\left(-\frac{1}{2}\right) - \frac{\sqrt{2}}{2}\cdot\frac{\sqrt{3}}{2} = \frac{-\sqrt{2}-\sqrt{6}}{4}.$

$\sin\frac{11\pi}{12} = \sin\left(\frac{\pi}{4} + \frac{2\pi}{3}\right) = \sin\frac{\pi}{4}\cos\frac{2\pi}{3} + \cos\frac{\pi}{4}\sin\frac{2\pi}{3} = \frac{\sqrt{2}}{2}\cdot\left(-\frac{1}{2}\right) + \frac{\sqrt{2}}{2}\cdot\frac{\sqrt{3}}{2} = \frac{-\sqrt{2}+\sqrt{6}}{4}.$

$\tan\frac{11\pi}{12} = \frac{\sin\frac{11\pi}{12}}{\cos\frac{11\pi}{12}} = \frac{-\sqrt{2}+\sqrt{6}}{4} \cdot \frac{4}{-\sqrt{2}-\sqrt{6}} = \frac{-\sqrt{2}+\sqrt{6}}{-\sqrt{2}-\sqrt{6}} = \frac{\sqrt{2}-\sqrt{6}}{\sqrt{2}+\sqrt{6}}.$

(iv) $\cos\frac{19\pi}{12} = \cos\left(\frac{5\pi}{4} + \frac{\pi}{3}\right) = \cos\frac{5\pi}{4}\cos\frac{\pi}{3} - \sin\frac{5\pi}{4}\sin\frac{\pi}{3} = -\frac{\sqrt{2}}{2}\cdot\frac{1}{2} - \left(-\frac{\sqrt{2}}{2}\right)\cdot\frac{\sqrt{3}}{2} = \frac{-\sqrt{2}+\sqrt{6}}{4}.$

$\sin\frac{19\pi}{12} = \sin\left(\frac{5\pi}{4} + \frac{\pi}{3}\right) = \sin\frac{5\pi}{4}\cos\frac{\pi}{3} + \cos\frac{5\pi}{4}\sin\frac{\pi}{3} = -\frac{\sqrt{2}}{2}\cdot\frac{1}{2} + \left(-\frac{\sqrt{2}}{2}\right)\cdot\frac{\sqrt{3}}{2} = \frac{-\sqrt{2}-\sqrt{6}}{4}.$

$\tan\frac{19\pi}{12} = \frac{\sin\frac{19\pi}{12}}{\cos\frac{19\pi}{12}} = \frac{-\sqrt{2}-\sqrt{6}}{4} \cdot \frac{4}{-\sqrt{2}+\sqrt{6}} = \frac{-\sqrt{2}-\sqrt{6}}{-\sqrt{2}+\sqrt{6}} = \frac{\sqrt{2}+\sqrt{6}}{\sqrt{2}-\sqrt{6}}.$

LEVEL 2

3. Each of the following complex numbers is written in exponential form. Rewrite each complex number in standard form: (i) $e^{\pi i}$; (ii) $e^{-\frac{5\pi}{2}i}$; (iii) $3e^{\frac{\pi}{4}i}$; (iv) $2e^{\frac{\pi}{3}i}$; (v) $\sqrt{2}e^{\frac{7\pi}{6}i}$; (vi) $\pi e^{-\frac{5\pi}{4}i}$; (vii) $e^{\frac{19\pi}{12}i}$

Solutions:

(i) $e^{\pi i} = \cos\pi + i\sin\pi = -1 + 0i = -\mathbf{1}.$

(ii) $e^{-\frac{5\pi}{2}i} = \cos\left(-\frac{5\pi}{2}\right) + i\sin\left(-\frac{5\pi}{2}\right) = \cos\frac{5\pi}{2} - i\sin\frac{5\pi}{2} = 0 - 1i = -\boldsymbol{i}.$

(iii) $3e^{\frac{\pi}{4}i} = 3\left(\cos\frac{\pi}{4} + i\sin\frac{\pi}{4}\right) = 3\left(\frac{\sqrt{2}}{2} + \frac{\sqrt{2}}{2}i\right) = \frac{3\sqrt{2}}{2} + \frac{3\sqrt{2}}{2}\boldsymbol{i}.$

(iv) $2e^{\frac{\pi}{3}i} = 2\left(\cos\frac{\pi}{3} + i\sin\frac{\pi}{3}\right) = 2\left(\frac{1}{2} + \frac{\sqrt{3}}{2}i\right) = 1 + \sqrt{3}\boldsymbol{i}.$

(v) $\sqrt{2}e^{\frac{7\pi}{6}i} = \sqrt{2}\left(\cos\frac{7\pi}{6} + i\sin\frac{7\pi}{6}\right) = \sqrt{2}\left(-\frac{\sqrt{3}}{2} - \frac{1}{2}i\right) = -\frac{\sqrt{6}}{2} - \frac{\sqrt{2}}{2}\boldsymbol{i}.$

(vi) $\pi e^{-\frac{5\pi}{4}i} = \pi\left(\cos\left(-\frac{5\pi}{4}\right) + i\sin\left(-\frac{5\pi}{4}\right)\right) = \pi\left(\cos\frac{5\pi}{4} - i\sin\frac{5\pi}{4}\right) = -\frac{\pi\sqrt{2}}{2} + \frac{\pi\sqrt{2}}{2}i.$

(vii) $e^{\frac{19\pi}{12}} = \cos\frac{19\pi}{12} + i\sin\frac{19\pi}{12} = \frac{-\sqrt{2}+\sqrt{6}}{4} + \frac{-\sqrt{2}-\sqrt{6}}{4}i$ (see part (iv) of Problem 2 above).

4. Each of the following complex numbers is written in standard form. Rewrite each complex number in exponential form: (i) $-1-i$; (ii) $\sqrt{3}+i$; (iii) $1-\sqrt{3}i$; (iv) $\left(\frac{\sqrt{6}+\sqrt{2}}{4}\right) + \left(\frac{\sqrt{6}-\sqrt{2}}{4}\right)i.$

Solutions:

(i) $r^2 = (-1)^2 + (-1)^2 = 1+1 = 2$. So, $r = \sqrt{2}$. $\tan\theta = \frac{-1}{-1} = 1$. So, $\theta = \pi + \frac{\pi}{4} = \frac{5\pi}{4}$. Therefore, $-1-i = \sqrt{2}e^{\frac{5\pi}{4}i} = \sqrt{2}e^{-\frac{3\pi}{4}i}.$

(ii) $r^2 = (\sqrt{3})^2 + 1^2 = 3+1 = 4$. So, $r = 2$. $\tan\theta = \frac{1}{\sqrt{3}}$. So, $\theta = \frac{\pi}{6}$. Therefore, we have $\sqrt{3}+i = 2e^{\frac{\pi}{6}i}.$

(iii) $r^2 = 1^2 + (-\sqrt{3})^2 = 1+3 = 4$. So, $r = 2$. $\tan\theta = \frac{-\sqrt{3}}{1}$. So, $\theta = -\frac{\pi}{3}$. Therefore, we have $1-\sqrt{3}i = 2e^{-\frac{\pi}{3}i}.$

(iv) $r^2 = \left(\frac{\sqrt{6}+\sqrt{2}}{4}\right)^2 + \left(\frac{\sqrt{6}-\sqrt{2}}{4}\right)^2 = \frac{6+2+2\sqrt{12}}{16} + \frac{6+2-2\sqrt{12}}{16} = \frac{16}{16} = 1$. So, $r = 1$. By part 2 of Problem 2, $\theta = \frac{\pi}{12}$. Therefore, $\left(\frac{\sqrt{6}+\sqrt{2}}{4}\right) + \left(\frac{\sqrt{6}-\sqrt{2}}{4}\right)i = 1e^{\frac{\pi}{12}i} = e^{\frac{\pi}{12}i}.$

5. Write the following complex numbers in standard form: (i) $\left(\frac{\sqrt{2}}{2} + \frac{\sqrt{2}}{2}i\right)^4$; (ii) $\left(1+\sqrt{3}i\right)^5.$

Solutions:

(i) If $z = \frac{\sqrt{2}}{2} + \frac{\sqrt{2}}{2}i$, then $r = \sqrt{\left(\frac{\sqrt{2}}{2}\right)^2 + \left(\frac{\sqrt{2}}{2}\right)^2} = \sqrt{\frac{2}{4} + \frac{2}{4}} = 1$ and $\tan\theta = \frac{\frac{\sqrt{2}}{2}}{\frac{\sqrt{2}}{2}} = 1$, so that $\theta = \frac{\pi}{4}$. So, in exponential form, $z = e^{\frac{\pi}{4}i}$. Therefore, $\left(\frac{\sqrt{2}}{2} + \frac{\sqrt{2}}{2}i\right)^4 = \left(e^{\frac{\pi}{4}i}\right)^4 = e^{\pi i} = -1.$

(ii) If $z = 1+\sqrt{3}i$, then $r = \sqrt{1^2 + (\sqrt{3})^2} = \sqrt{1+3} = \sqrt{4} = 2$ and $\tan\theta = \frac{\sqrt{3}}{1} = \sqrt{3}$, so that $\theta = \frac{\pi}{3}$. So, in exponential form, $z = 2e^{\frac{\pi}{3}i}$. So, $\left(1+\sqrt{3}i\right)^5 = \left(2e^{\frac{\pi}{3}i}\right)^5 = 2^5 e^{\frac{5\pi}{3}i} = 16 - 16\sqrt{3}i.$

LEVEL 3

6. Use De Moivre's Theorem to prove the following identities: (i) $\cos 2\theta = \cos^2\theta - \sin^2\theta$; (ii) $\sin 2\theta = 2\sin\theta\cos\theta$; (iii) $\cos 3\theta = \cos^3\theta - 3\cos\theta\sin^2\theta.$

Proofs:

(i) By De Moivre's Theorem, $\left(e^{i\theta}\right)^2 = e^{i(2\theta)}$, so that $(\cos\theta + i\sin\theta)^2 = \cos 2\theta + i\sin 2\theta.$ Multiplying the left-hand side gives us

$$(\cos\theta + i\sin\theta)^2 = (\cos\theta + i\sin\theta)(\cos\theta + i\sin\theta) = \cos^2\theta - \sin^2\theta + i(2\sin\theta\cos\theta).$$

So, $\cos^2\theta - \sin^2\theta + i(2\sin\theta\cos\theta) = \cos 2\theta + i\sin 2\theta$. Equating the real parts of this equation gives us $\cos^2\theta - \sin^2\theta = \cos 2\theta$. □

(ii) By (i), we have $\cos^2\theta - \sin^2\theta + i(2\sin\theta\cos\theta) = \cos 2\theta + i\sin 2\theta$. Equating the imaginary parts of this equation gives us $2\sin\theta\cos\theta = \sin 2\theta$. □

(iii) By De Moivre's Theorem, $\left(e^{i\theta}\right)^3 = e^{i(3\theta)}$, so that $(\cos\theta + i\sin\theta)^3 = \cos 3\theta + i\sin 3\theta$. Multiplying the left-hand side and using the computation from (i) gives us

$$(\cos\theta + i\sin\theta)^3 = \left(\cos^2\theta - \sin^2\theta + i(2\sin\theta\cos\theta)\right)(\cos\theta + i\sin\theta)$$
$$= (\cos^3\theta - \cos\theta\sin^2\theta - 2\cos\theta\sin^2\theta) + i(\cos^2\theta\sin\theta - \sin^3\theta + 2\cos^2\theta\sin\theta)$$
$$= (\cos^3\theta - 3\cos\theta\sin^2\theta) + i(3\cos^2\theta\sin\theta - \sin^3\theta)$$

So, $(\cos^3\theta - 3\cos\theta\sin^2\theta) + i(3\cos^2\theta\sin\theta - \sin^3\theta) = \cos 3\theta + i\sin 3\theta$. Equating the real parts of this equation gives us $\cos^3\theta - 3\cos\theta\sin^2\theta = \cos 3\theta$. □

Note: Equating imaginary parts in (iii) gives us one more trigonometric identity:
$$\sin 3\theta = 3\cos^2\theta\sin\theta - \sin^3\theta$$

7. Suppose that $z = re^{i\theta}$ and $w = se^{i\phi}$ are complex numbers written in exponential form. Express each of the following in exponential form. Provide a proof in each case: (i) zw; (ii) $\frac{z}{w}$.

Solutions:

(i) $zw = re^{i\theta}se^{i\phi} = rse^{i\theta}e^{i\phi} = rs(\cos\theta + i\sin\theta)(\cos\phi + i\sin\phi)$
$= rs[(\cos\theta\cos\phi - \sin\theta\sin\phi) + i(\sin\theta\cos\phi + \cos\theta\sin\phi)]$
$= rs[\cos(\theta + \phi) + i\sin(\theta + \phi)] = \boldsymbol{rse^{(\theta+\phi)i}}$.

(ii) $\frac{z}{w} = \frac{re^{i\theta}}{se^{i\phi}} = \frac{r}{s}\cdot\frac{e^{i\theta}e^{-i\phi}}{e^{i\phi}e^{-i\phi}} = \frac{r}{s}\cdot\frac{e^{i(\theta-\phi)}}{e^{i(\phi-\phi)}} = \frac{r}{s}e^{i(\theta-\phi)}$.

8. Write each function in the form $f(z) = u(x, y) + iv(x, y)$ and $f(z) = u(r, \theta) + iv(r, \theta)$:
(i) $f(z) = 2z^2 - 5$; (ii) $f(z) = \frac{1}{z}$; (iii) $f(z) = z^3 + z^2 + z + 1$.

Solutions:

(i) $f(z) = 2z^2 - 5 = 2(x + yi)^2 - 5 = 2(x^2 - y^2 + 2xyi) - 5 = (\boldsymbol{2x^2 - 2y^2 - 5}) + \boldsymbol{4xyi}$

$f(z) = 2z^2 - 5 = 2\left(re^{i\theta}\right)^2 - 5 = 2\left(r^2e^{i(2\theta)}\right) - 5 = 2r^2e^{i(2\theta)} - 5$
$= 2r^2(\cos 2\theta + i\sin 2\theta) - 5 = (\boldsymbol{2r^2\cos(2\theta) - 5}) + i(\boldsymbol{2r^2\sin 2\theta})$

(ii) $f(z) = \frac{1}{z} = \frac{1}{z}\cdot\frac{\bar{z}}{\bar{z}} = \frac{\bar{z}}{z\bar{z}} = \frac{x-yi}{x^2+y^2} = \frac{x}{x^2+y^2} - \frac{y}{x^2+y^2}i$

$f(z) = \frac{1}{z} = \frac{1}{re^{i\theta}} = \frac{1}{r}e^{-i\theta} = \frac{1}{r}(\cos(-\theta) + i\sin(-\theta)) = \frac{1}{r}(\cos\theta - i\sin\theta)$
$= \frac{1}{r}\cos\theta + i\left(\frac{1}{r}\sin\theta\right)$

(iii) $f(z) = z^3 + z^2 + z + 1 = (x + yi)(x^2 - y^2 + 2xyi) + (x^2 - y^2 + 2xyi) + (x + yi) + 1$
$= x^3 - xy^2 - 2xy^2 + (x^2y - y^3 + 2x^2y)i + (x^2 - y^2 + x + 1) + (2xy + y)i$
$= (x^3 - 3xy^2) + (3x^2y - y^3)i + (x^2 - y^2 + x + 1) + (2xy + y)i$
$= (x^3 - 3xy^2 + x^2 - y^2 + x + 1) + (3x^2y - y^3 + 2xy + y)i$

$f(z) = z^3 + z^2 + z + 1 = (re^{i\theta})^3 + (re^{i\theta})^2 + re^{i\theta} + 1 = r^3 e^{i(3\theta)} + r^2 e^{i(2\theta)} + re^{i\theta} + 1$
$= r^3(\cos 3\theta + i \sin 3\theta) + r^2(\cos 2\theta + i \sin 2\theta) + r(\cos\theta + i \sin\theta) + 1$
$= (r^3 \cos(3\theta) + r^2 \cos(2\theta) + r \cos\theta + 1) + i(r^3 \sin(3\theta) + r^2 \sin(2\theta) + r \sin\theta)$

9. Let $f(z) = x^2 - y^2 - 2x + 2y(x + 1)i$. Rewrite $f(z)$ in terms of z.

Solution: $f(z) = x^2 - y^2 - 2x + 2y(x+1)i = x^2 - y^2 + 2xyi - 2(x - yi) = z^2 - 2\bar{z}$.

Note: We can also do this formally by replacing $x = \text{Re } z$ by $\frac{z+\bar{z}}{2}$ and $y = \text{Im } z$ by $\frac{z-\bar{z}}{2i}$. In this case, we get

$$f(z) = \left(\frac{z+\bar{z}}{2}\right)^2 - \left(\frac{z-\bar{z}}{2i}\right)^2 - 2\left(\frac{z+\bar{z}}{2}\right) + 2\left(\frac{z-\bar{z}}{2i}\right)\left(\frac{z+\bar{z}}{2} + 1\right)i$$

$$= \frac{z^2 + 2z\bar{z} + \bar{z}^2}{4} - \frac{z^2 - 2z\bar{z} + \bar{z}^2}{-4} - (z + \bar{z}) + (z - \bar{z})\left(\frac{z+\bar{z}}{2} + 1\right)$$

$$= \frac{z^2 + 2z\bar{z} + \bar{z}^2}{4} + \frac{z^2 - 2z\bar{z} + \bar{z}^2}{4} - z - \bar{z} + (z - \bar{z})\left(\frac{z+\bar{z}}{2}\right) + (z - \bar{z})$$

$$= \frac{2z^2 + 2\bar{z}^2}{4} - z - \bar{z} + \frac{1}{2}(z^2 - \bar{z}^2) + z - \bar{z}$$

$$= \frac{1}{2}(z^2 + \bar{z}^2) - z - \bar{z} + \frac{1}{2}(z^2 - \bar{z}^2) + z - \bar{z}$$

$$= \frac{1}{2}(z^2 + \bar{z}^2) + \frac{1}{2}(z^2 - \bar{z}^2) - 2\bar{z}$$

$$= z^2 - 2\bar{z}.$$

10. Find all complex numbers that satisfy the given equation: (i) $z^6 - 1 = 0$; (ii) $z^4 + 4 = 0$.

Solutions:

(i) We are looking for the sixth roots of unity. So, $z = \sqrt[6]{1}e^{i\left(\frac{0}{6} + \frac{2k\pi}{6}\right)} = e^{\frac{k\pi}{3}i}$ for $k = 0, 1, 2, 3, 4, 5$. Substituting each of these values for k into the expression $e^{\frac{k\pi}{3}i}$ gives us the following 6 sixth roots of unity.

$$1, \frac{\sqrt{3}}{2} + \frac{1}{2}i, -\frac{\sqrt{3}}{2} + \frac{1}{2}i, -1, -\frac{\sqrt{3}}{2} - \frac{1}{2}, \frac{\sqrt{3}}{2} - \frac{1}{2}i$$

(ii) We are looking for the fourth roots of -4. So, $z = \sqrt[4]{4}e^{i\left(\frac{\pi}{4}+\frac{2k\pi}{4}\right)} = \sqrt{2}e^{\frac{(2k+1)\pi}{4}i}$ for $k = 0, 1, 2, 3$. Substituting each of these values for k into the expression $e^{\frac{k\pi}{3}i}$ gives us the following 4 fourth roots of -4.

$$1+i, -1+i, -1-i, 1-i$$

LEVEL 4

11. Consider triangle AOP, where $O = (0,0)$, $A = (1,0)$, and P is the point on the unit circle so that angle POA has radian measure $\frac{\pi}{3}$. Prove that triangle AOP is equilateral, and then use this to prove that $W\left(\frac{\pi}{3}\right) = \left(\frac{1}{2}, \frac{\sqrt{3}}{2}\right)$. You may use the following facts about triangles: (i) The interior angle measures of a triangle sum to π radians; (ii) Two sides of a triangle have the same length if and only if the interior angles of the triangle opposite these sides have the same measure; (iii) If two sides of a triangle have the same length, then the line segment beginning at the point of intersection of those two sides and terminating on the opposite base midway between the endpoints of that base is perpendicular to that base.

Proof: Let's start by drawing the unit circle together with triangle AOP. We also draw line segment PE, where E is midway between O and A.

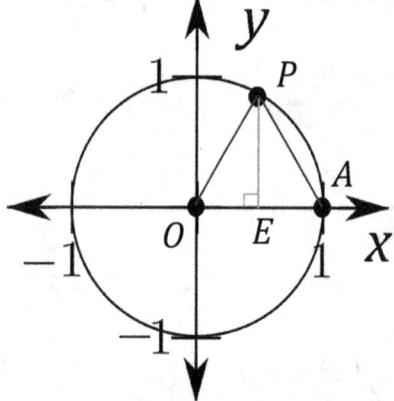

Since OP and OA are both radii of the circle, they have the same length. By (ii), angles OAP and OPA have the same measure. By (i), the sum of these measures is $\pi - \frac{\pi}{3} = \frac{3\pi}{3} - \frac{\pi}{3} = \frac{2\pi}{3}$. So, each of angles OAP and OPA measure $\frac{\pi}{3}$ radians. It follows from (ii) again that triangle AOP is equilateral.

By (iii), PE is perpendicular to OA.

Now, $OP = 1$ because OP is a radius of the unit circle and $OE = \frac{1}{2}$ because OA is a radius of the unit circle and E is midway between O and A. Since triangle OEP is a right triangle with hypotenuse OP, by the Pythagorean Theorem, $PE^2 = OP^2 - OE^2 = 1^2 - \left(\frac{1}{2}\right)^2 = 1 - \frac{1}{4} = \frac{3}{4}$. So, $PE = \sqrt{\frac{3}{4}} = \frac{\sqrt{3}}{\sqrt{4}} = \frac{\sqrt{3}}{2}$. It follows that $W\left(\frac{\pi}{3}\right) = \left(\frac{1}{2}, \frac{\sqrt{3}}{2}\right)$. □

12. Prove that $W\left(\frac{\pi}{6}\right) = \left(\frac{\sqrt{3}}{2}, \frac{1}{2}\right)$. You can use facts (i), (ii), and (iii) described in Problem 11.

Proof: Let's start by drawing a picture similar to what we drew in Problem 1. We draw P and Q on the unit circle and A on the positive x-axis so that angle AOP has radian measure $\frac{\pi}{6}$, angle AOQ has radian measure $-\frac{\pi}{6}$, and A is right in the middle of the line segment joining P and Q.

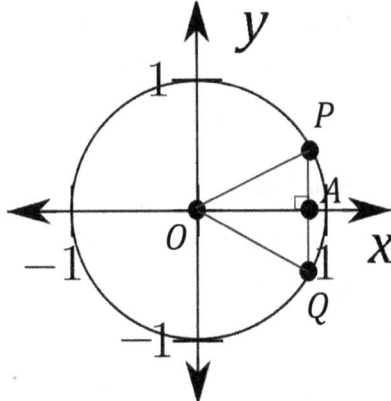

By reasoning similar to what was done in Problem 1, we see that triangle POQ is equilateral and OA is perpendicular to PQ.

Now, $OP = 1$ because OP is a radius of the unit circle and $AP = \frac{1}{2}$ because A is midway between P and Q. Since triangle POA is a right triangle with hypotenuse OP, by the Pythagorean Theorem, $OA^2 = OP^2 - AP^2 = 1^2 - \left(\frac{1}{2}\right)^2 = 1 - \frac{1}{4} = \frac{3}{4}$. Therefore, $OA = \sqrt{\frac{3}{4}} = \frac{\sqrt{3}}{\sqrt{4}} = \frac{\sqrt{3}}{2}$. It follows that $W\left(\frac{\pi}{6}\right) = \left(\frac{\sqrt{3}}{2}, \frac{1}{2}\right)$. □

13. Let θ and ϕ be the radian measure of angles A and B, respectively. Prove the following identity:
$$\cos(\theta - \phi) = \cos\theta \cos\phi + \sin\theta \sin\phi$$

Proof: Let's draw a picture of the unit circle together with angles θ, ϕ, and $\theta - \phi$ in standard position, and label the corresponding points on the unit circle.

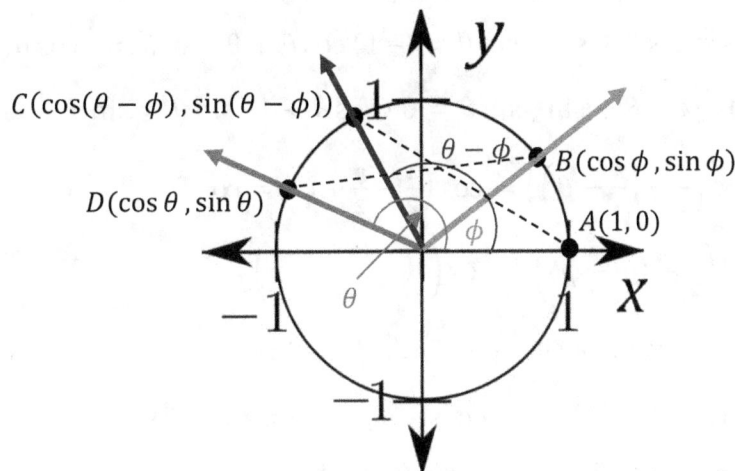

Since the arcs moving counterclockwise from A to C and from B to D both have radian measure $\theta - \phi$, it follows that $AC = BD$, and so, using the Pythagorean Theorem twice, we have
$$(\cos(\theta - \phi) - 1)^2 + (\sin(\theta - \phi) - 0)^2 = (\cos\theta - \cos\phi)^2 + (\sin\theta - \sin\phi)^2$$

The left-hand side of this equation is equal to:

$$(\cos(\theta - \phi) - 1)^2 + (\sin(\theta - \phi) - 0)^2$$
$$= \cos^2(\theta - \phi) - 2\cos(\theta - \phi) + 1 + \sin^2(\theta - \phi)$$
$$= (\cos^2(\theta - \phi) + \sin^2(\theta - \phi)) - 2\cos(\theta - \phi) + 1$$
$$= 1 - 2\cos(\theta - \phi) + 1 \text{ (by the Pythagorean Identity)}$$
$$= 2 - 2\cos(\theta - \phi)$$

The right-hand side of this equation is equal to:

$$(\cos\theta - \cos\phi)^2 + (\sin\theta - \sin\phi)^2$$
$$= \cos^2\theta - 2\cos\theta\cos\phi + \cos^2\phi + \sin^2\theta - 2\sin\theta\sin\phi + \sin^2\phi$$
$$= (\cos^2\theta + \sin^2\theta) + (\cos^2\phi + \sin^2\phi) - 2\cos\theta\cos\phi - 2\sin\theta\sin\phi$$
$$= 1 + 1 - 2\cos\theta\cos\phi - 2\sin\theta\sin\phi$$
$$= 2 - 2\cos\theta\cos\phi - 2\sin\theta\sin\phi$$

Therefore, we have $2 - 2\cos(\theta - \phi) = 2 - 2\cos\theta\cos\phi - 2\sin\theta\sin\phi$. Subtracting 2 from each side of this equation gives us $-2\cos(\theta - \phi) = -2\cos\theta\cos\phi - 2\sin\theta\sin\phi$. Multiplying each side of this last equation by $-\frac{1}{2}$ gives us $\cos(\theta - \phi) = \cos\theta\cos\phi + \sin\theta\sin\phi$, as desired. □

14. Let θ and ϕ be the radian measure of angles A and B, respectively. Prove the following identities:
(i) $\cos(\theta + \phi) = \cos\theta\cos\phi - \sin\theta\sin\phi$; (ii) $\cos(\pi - \theta) = -\cos\theta$; (iii) $\cos\left(\frac{\pi}{2} - \theta\right) = \sin\theta$;
(iv) $\sin\left(\frac{\pi}{2} - \theta\right) = \cos\theta$; (v) $\sin(\theta + \phi) = \sin\theta\cos\phi + \cos\theta\sin\phi$; (vi) $\sin(\pi - \theta) = \sin\theta$.

Proofs:

(i) $\cos(\theta + \phi) = \cos(\theta - (-\phi)) = \cos\theta\cos(-\phi) + \sin\theta\sin(-\phi)$ (by Problem 13)
$$= \cos\theta\cos\phi - \sin\theta\sin\phi \text{ (by the Negative Identities).} \qquad \square$$

(ii) $\cos(\pi - \theta) = \cos\pi\cos\theta + \sin\pi\sin\theta = (-1)\cos\theta + 0\cdot\sin\theta = -\cos\theta.$ □

(iii) $\cos\left(\frac{\pi}{2} - \theta\right) = \cos\frac{\pi}{2}\cos\theta + \sin\frac{\pi}{2}\sin\theta = 0\cdot\cos\theta + 1\cdot\sin\theta = \sin\theta.$ □

(iv) $\sin\left(\frac{\pi}{2} - \theta\right) = \cos\left(\frac{\pi}{2} - \left(\frac{\pi}{2} - \theta\right)\right) = \cos\left(\frac{\pi}{2} - \frac{\pi}{2} + \theta\right) = \cos\theta.$ □

(v) $\sin(\theta + \phi) = \cos\left(\frac{\pi}{2} - (\theta + \phi)\right) = \cos\left(\left(\frac{\pi}{2} - \theta\right) - \phi\right)$
$$= \cos\left(\frac{\pi}{2} - \theta\right)\cos\phi + \sin\left(\frac{\pi}{2} - \theta\right)\sin\phi = \sin\theta\cos\phi + \cos\theta\sin\phi. \qquad \square$$

(vi) $\sin(\pi - \theta) = \sin(\pi + (-\theta)) = \sin\pi\cos(-\theta) + \cos\pi\sin(-\theta)$
$$= 0\cdot\cos\theta + (-1)(-\sin\theta) = \sin\theta. \qquad \square$$

15. Let $z, w \in \mathbb{C}$. Prove that $\arg zw = \arg z + \arg w$ in the sense that if two of the three terms in the equation are specified, then there is a value for the third term so that the equation holds. Similarly, prove that $\arg \frac{z}{w} = \arg z - \arg w$. Finally, provide examples to show that the corresponding equations are false if we replace "arg" by "Arg."

Proof: Let θ and ϕ be any values of $\arg z$ and $\arg w$, respectively. Then there are positive real numbers r and s such that $z = re^{i\theta}$ and $w = se^{i\phi}$. By the solution to Problem 7, $zw = rse^{i(\theta+\phi)}$. So, $\theta + \phi$ is a value of $\arg zw$. Next, without loss of generality, choose values for $\arg zw$ and $\arg z$, so that for some $j, k \in \mathbb{Z}$, $\arg zw = (\theta + \phi) + 2j\pi$ and $\arg z = \theta + 2k\pi$. If we use $\phi + 2(j - k)\pi$ for $\arg w$, we have

$$\arg zw = (\theta + \phi) + 2j\pi = (\theta + 2k\pi) + (\phi + 2(j - k)\pi) = \arg z + \arg w.$$

Once again, let θ and ϕ be any values of $\arg z$ and $\arg w$, respectively. Then there are positive real numbers r and s such that $z = re^{i\theta}$ and $w = se^{i\phi}$. By the solution to Problem 7, $\frac{z}{w} = \frac{r}{s}e^{i(\theta-\phi)}$. So, $\theta - \phi$ is a value of $\arg \frac{z}{w}$. Finally, without loss of generality, choose values for $\arg \frac{z}{w}$ and $\arg z$, so that for some $j, k \in \mathbb{Z}$, $\arg \frac{z}{w} = (\theta - \phi) + 2j\pi$ and $\arg z = \theta + 2k\pi$. If we use $\phi + 2(k - j)\pi$ for $\arg w$, we have

$$\arg \frac{z}{w} = (\theta - \phi) + 2j\pi = (\theta + 2k\pi) - (\phi + 2(k - j)\pi) = \arg z - \arg w.$$

To see that the equation $\operatorname{Arg} zw = \operatorname{Arg} z + \operatorname{Arg} w$ can be false, let $z = e^{\frac{3\pi}{4}i}$ and $w = e^{\frac{\pi}{2}i}$. Then we have $zw = e^{\frac{3\pi}{4}i}e^{\frac{\pi}{2}i} = e^{(\frac{3\pi}{4}+\frac{\pi}{2})i} = e^{\frac{5\pi}{4}i} = e^{-\frac{3\pi}{4}i}$. So, $\operatorname{Arg} zw = -\frac{3\pi}{4}$, whereas $\operatorname{Arg} z + \operatorname{Arg} w = \frac{3\pi}{4} + \frac{\pi}{2} = \frac{5\pi}{4}$.

To see that the equation $\operatorname{Arg} \frac{z}{w} = \operatorname{Arg} z - \operatorname{Arg} w$ can be false, let $z = e^{\frac{3\pi}{4}i}$ and $w = e^{-\frac{\pi}{2}i}$. Then we have $\frac{z}{w} = \frac{e^{\frac{3\pi}{4}i}}{e^{-\frac{\pi}{2}i}} = e^{(\frac{3\pi}{4}+\frac{\pi}{2})i} = e^{\frac{5\pi}{4}i} = e^{-\frac{3\pi}{4}i}$. So, $\operatorname{Arg} \frac{z}{w} = -\frac{3\pi}{4}$, whereas $\operatorname{Arg} z - \operatorname{Arg} w = \frac{3\pi}{4} - (-\frac{\pi}{2}) = \frac{5\pi}{4}$. □

LEVEL 5

16. Define the function $f: \mathbb{C} \to \mathbb{C}$ by $f(z) = z^2$. Determine the images under f of each of the following sets: (i) $A = \{x + yi \mid x^2 - y^2 = 1\}$; (ii) $B = \{x + yi \mid x > 0 \land y > 0 \land xy < 1\}$; (iii) $C = \{x + yi \mid x \geq 0 \land y \geq 0\}$; (vi) $D = \{x + yi \mid y \geq 0\}$.

Solutions:

(i) $f(z) = z^2 = x^2 - y^2 + 2xyi$. So, $u(x, y) = x^2 - y^2$ and $v(x, y) = 2xy$. When $x^2 - y^2 = 1$, $u(x, y) = 1$. It follows that $f[A] \subseteq \{u + vi \mid u = 1\}$. Now, $x^2 - y^2 = 1$ is equivalent to $x^2 = y^2 + 1$ or $x = \pm\sqrt{y^2 + 1}$. When $x = \sqrt{y^2 + 1}$, $v(x, y) = 2y\sqrt{y^2 + 1}$. v is a continuous function with $\lim_{y \to -\infty} v(x, y) = -\infty$ and $\lim_{y \to +\infty} v(x, y) = +\infty$. It follows that the image of A under the function f defined by $f(z) = z^2$ is the entire vertical line $u = 1$. In other words, we have $f[A] = \{u + vi \mid u = 1\}$.

(ii) When $0 < xy < 1$, $0 < v(x,y) < 2$. It follows that $f[B] \subseteq \{u+vi \mid 0 < v < 2\}$. Let's choose an arbitrary but specific real number a between 0 and 2 and consider $v(x,y) = a$. Then we have $2xy = a$, or equivalently, $y = \frac{a}{2x}$. So, $u(x,y) = x^2 - \left(\frac{a}{2x}\right)^2 = x^2 - \frac{a^2}{4x^2}$. Now, u is continuous on $(0, +\infty)$ with $\lim_{x \to 0^+} u(x,y) = -\infty$ and $\lim_{x \to +\infty} u(x,y) = +\infty$. It follows that the image of B under the function f defined by $f(z) = z^2$ is the entire horizontal strip $0 < v < 2$. In other words, we have $f[B] = \{u+vi \mid 0 < v < 2\}$.

(iii) As we saw in part 4 of Example 15.7, $f(z) = r^2 e^{i(2\theta)} = r^2(\cos 2\theta + i \sin 2\theta)$. So, $u(r,\theta) = r^2 \cos 2\theta$ and $v(r,\theta) = r^2 \sin 2\theta$. Now, if $x \geq 0$ and $y \geq 0$, then $0 \leq \theta \leq \frac{\pi}{2}$. It follows that $0 \leq 2\theta \leq \pi$. So, $0 \leq \sin 2\theta \leq 1$, and therefore, $0 \leq r^2 \sin 2\theta \leq r^2$ as r ranges over all possible nonnegative real numbers. Thus, $v \geq 0$. So, $f[C] \subseteq \{u+vi \mid v \geq 0\}$. Now, if $u + vi$ is an arbitrary complex number with $v \geq 0$, let $u = r \cos \phi$ and $v = r \sin \phi$ with $r \geq 0$ and $0 \leq \phi \leq \pi$. Let $z = \sqrt{r} e^{\frac{\phi}{2}i}$. Then $z \in C$ and $f(z) = re^{\phi i}$. It follows that the image under C is the entire half plane $v \geq 0$. In other words, we have $f[C] = \{u+vi \mid v \geq 0\}$.

(iv) Once again, we have $f(z) = r^2 e^{i(2\theta)} = r^2(\cos 2\theta + i \sin 2\theta)$, so that $u(r,\theta) = r^2 \cos 2\theta$ and $v(r,\theta) = r^2 \sin 2\theta$. Now, if $y \geq 0$, then $0 \leq \theta \leq \pi$. We will show that $f[D] = \mathbb{C}$. If $u + vi$ is an arbitrary complex number, let $u = r \cos \phi$ and $v = r \sin \phi$ with $r \geq 0$ and $0 \leq \phi < 2\pi$. Let $z = \sqrt{r} e^{\frac{\phi}{2}i}$. Then $z \in D$ and $f(z) = re^{\phi i}$. It follows that the image under D is the entire complex plane. In other words, we have $f[D] = \mathbb{C}$.

17. Let $A \subseteq \mathbb{C}$, let $f: A \to \mathbb{C}$, let $L = j + ki \in \mathbb{C}$, and let $a = b + ci \in \mathbb{C}$ be a point such that A contains some deleted neighborhood of a. Suppose that $f(x + yi) = u(x,y) + iv(x,y)$. Prove that $\lim_{z \to a} f(z) = L$ if and only if $\lim_{(x,y) \to (b,c)} u(x,y) = j$ and $\lim_{(x,y) \to (b,c)} v(x,y) = k$.

Proof: Suppose that $\lim_{z \to a} f(z) = L$ and let $\epsilon > 0$. Then there is $\delta > 0$ such that $0 < |z - a| < \delta$ implies $|f(z) - L| < \epsilon$. Now,

$$|z - a| = |(x + yi) - (b + ci)| = |(x - b) + (y - c)i| = \sqrt{(x-b)^2 + (y-c)^2}$$

Also, $|f(z) - L| = |(u(x,y) + iv(x,y)) - (j + ki)| = |(u(x,y) - j) + (v(x,y) - k)i|$.

So, if $0 < \sqrt{(x-b)^2 + (y-c)^2} < \delta$, then $0 < |z - a| < \delta$, and therefore, $|f(z) - L| < \epsilon$. It follows that

$$|u(x,y) - j| \leq |(u(x,y) - j) + (v(x,y) - k)i| = |f(z) - L| < \epsilon$$

and

$$|v(x,y) - k| \leq |(u(x,y) - j) + (v(x,y) - k)i| = |f(z) - L| < \epsilon$$

Therefore, $\lim_{(x,y) \to (b,c)} u(x,y) = j$ and $\lim_{(x,y) \to (b,c)} v(x,y) = k$.

Conversely, suppose that $\lim_{(x,y) \to (b,c)} u(x,y) = j$ and $\lim_{(x,y) \to (b,c)} v(x,y) = k$ and let $\epsilon > 0$. Then there are $\delta_1, \delta_2 > 0$ such that

$$0 < \sqrt{(x-b)^2 + (y-c)^2} < \delta_1 \text{ implies } |u(x,y) - j| < \frac{\epsilon}{2}$$

and

$$0 < \sqrt{(x-b)^2 + (y-c)^2} < \delta_2 \text{ implies } |v(x,y) - k| < \frac{\epsilon}{2}.$$

Let $\delta = \min\{\delta_1, \delta_2\}$ and assume that $0 < |z - a| < \delta$. Since $|z - a| = \sqrt{(x-b)^2 + (y-c)^2}$ and $\delta \leq \delta_1, \delta_2$, we have $|u(x,y) - j| < \frac{\epsilon}{2}$ and $|v(x,y) - k| < \frac{\epsilon}{2}$. It follows that

$$|f(z) - L| = |(u(x,y) + iv(x,y)) - (j + ki)| = |(u(x,y) - j) + (v(x,y) - k)i|$$
$$\leq |u(x,y) - j| + |v(x,y) - k| < \frac{\epsilon}{2} + \frac{\epsilon}{2} = \epsilon.$$

Therefore, $\lim_{z \to a} f(z) = L$. □

18. Give a reasonable definition for each of the following limits (like what was done right before Theorem 15.4). L is a finite real number. (i) $\lim_{z \to \infty} f(z) = L$; (ii) $\lim_{z \to \infty} f(z) = \infty$.

Equivalent definitions:

(i) $\lim_{z \to \infty} f(z) = L$ if and only if $\forall \epsilon > 0 \, \exists \delta > 0 \left(|z| > \frac{1}{\delta} \to |f(z) - L| < \epsilon\right)$.

(ii) $\lim_{z \to \infty} f(z) = \infty$ if and only if $\forall \epsilon > 0 \, \exists \delta > 0 \left(|z| > \frac{1}{\delta} \to |f(z)| > \frac{1}{\epsilon}\right)$.

19. Prove each of the following: (i) $\lim_{z \to \infty} f(z) = L$ if and only $\lim_{z \to 0} f\left(\frac{1}{z}\right) = L$; (ii) $\lim_{z \to \infty} f(z) = \infty$ if and only $\lim_{z \to 0} \frac{1}{f\left(\frac{1}{z}\right)} = 0$.

Proofs:

(i) Suppose that $\lim_{z \to \infty} f(z) = L$ and let $\epsilon > 0$. There is $\delta > 0$ so that $|z| > \frac{1}{\delta} \to |f(z) - L| < \epsilon$. If we let $w = \frac{1}{z}$, we have $z = \frac{1}{w}$, and therefore, $\left|\frac{1}{w}\right| > \frac{1}{\delta} \to \left|f\left(\frac{1}{w}\right) - L\right| < \epsilon$. But $\left|\frac{1}{w}\right| > \frac{1}{\delta}$ is equivalent to $0 < |w| < \delta$. So, $0 < |w - 0| < \delta \to \left|f\left(\frac{1}{w}\right) - L\right| < \epsilon$. Thus, $\lim_{w \to 0} f\left(\frac{1}{w}\right) = L$. This is equivalent to $\lim_{z \to 0} f\left(\frac{1}{z}\right) = L$.

Conversely, suppose that $\lim_{z \to 0} f\left(\frac{1}{z}\right) = L$ and let $\epsilon > 0$. Then there is $\delta > 0$ so that $0 < |z - 0| < \delta \to \left|f\left(\frac{1}{z}\right) - L\right| < \epsilon$. If we let $w = \frac{1}{z}$, then $z = \frac{1}{w}$, and therefore, we have $0 < \left|\frac{1}{w}\right| < \delta \to |f(w) - L| < \epsilon$. Now, $0 < \left|\frac{1}{w}\right| < \delta$ is equivalent to $|w| > \frac{1}{\delta}$. So, we have $|w| > \frac{1}{\delta} \to |f(w) - L| < \epsilon$. Therefore, $\lim_{w \to \infty} f(w) = L$, or equivalently, $\lim_{z \to \infty} f(z) = L$. □

(ii) Suppose that $\lim_{z \to \infty} f(z) = \infty$ and let $\epsilon > 0$. There is $\delta > 0$ so that $|z| > \frac{1}{\delta} \to |f(z)| > \frac{1}{\epsilon}$. If we let $w = \frac{1}{z}$, we have $z = \frac{1}{w}$, and therefore, $\left|\frac{1}{w}\right| > \frac{1}{\delta} \to \left|f\left(\frac{1}{w}\right)\right| > \frac{1}{\epsilon}$. But, $\left|\frac{1}{w}\right| > \frac{1}{\delta}$ is equivalent to $0 < |w| < \delta$ and $\left|f\left(\frac{1}{w}\right)\right| > \frac{1}{\epsilon}$ is equivalent to $\left|\frac{1}{f\left(\frac{1}{w}\right)}\right| < \epsilon$. So, $0 < |w| < \delta \to \left|\frac{1}{f\left(\frac{1}{w}\right)}\right| < \epsilon$, and therefore, $\lim_{w \to 0} \frac{1}{f\left(\frac{1}{w}\right)} = 0$. This is equivalent to $\lim_{z \to 0} \frac{1}{f\left(\frac{1}{z}\right)} = 0$.

Now, let $\lim_{z \to 0} \frac{1}{f\left(\frac{1}{z}\right)} = 0$ and let $\epsilon > 0$. There is $\delta > 0$ so that $0 < |z - 0| < \delta \to \left|\frac{1}{f\left(\frac{1}{z}\right)} - 0\right| < \epsilon$. If we let $w = \frac{1}{z}$, we have $z = \frac{1}{w}$, and therefore, $0 < |z - 0| < \delta$ is equivalent to $0 < \left|\frac{1}{w}\right| < \delta$, or equivalently, $|w| > \frac{1}{\delta}$. Also, $\left|\frac{1}{f\left(\frac{1}{z}\right)} - 0\right| < \epsilon$ is equivalent to $\left|\frac{1}{f(w)}\right| < \epsilon$, which in turn is equivalent to $|f(w)| > \frac{1}{\epsilon}$. So, $|w| > \frac{1}{\delta} \to |f(w)| > \frac{1}{\epsilon}$. Thus, $\lim_{w \to \infty} f(w) = \infty$. This is equivalent to $\lim_{z \to \infty} f(z) = \infty$. □

20. Let $f, g: \mathbb{R} \to \mathbb{R}$ be defined by $f(x) = \cos x$ and $g(x) = \sin x$. Prove that f and g are uniformly continuous on $\mathbb{R}$. Hint: Use the fact that the least distance between two points is a straight line.

Proof: Let $\epsilon > 0$ and let $\delta = \min\{\epsilon, 2\pi\}$. Let $x, y \in \mathbb{R}$ with $|x - y| < \delta$. Suppose that $W(x) = (a, b)$ and $W(y) = (c, d)$. The arc length along the unit circle between (a, b) and (c, d) is $|x - y|$ and the straight-line distance between (a, b) and (c, d) is $\sqrt{(a - c)^2 + (b - d)^2}$. Thus,

$$|\cos x - \cos y| = |a - c| \leq \sqrt{(a - c)^2 + (b - d)^2}$$
$$\leq |x - y|.$$

$$|\sin x - \sin y| = |b - d| \leq \sqrt{(a - c)^2 + (b - d)^2}$$
$$\leq |x - y|.$$

Therefore, we have $|\cos x - \cos y| \leq |x - y| < \delta \leq \epsilon$ and $|\sin x - \sin y| \leq |x - y| < \delta \leq \epsilon$. It follows that f and g are uniformly continuous on $\mathbb{R}$.

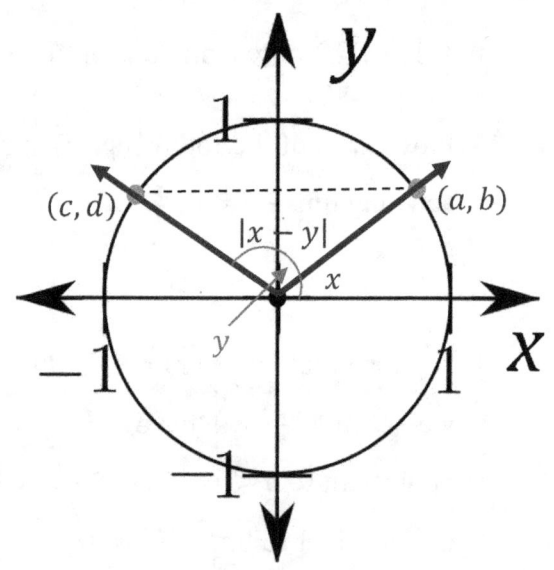

CHALLENGE PROBLEM

21. Consider $\mathbb{C}$ with the standard topology and $\mathbb{S}^2$ with its subspace topology, where $\mathbb{S}^2$ is being considered as a subspace of $\mathbb{R}^3$. Let $f: \mathbb{C} \to \mathbb{S}^2 \setminus N$ be defined as follows:

$$f(z) = \left(\frac{z + \bar{z}}{1 + |z|^2}, \frac{z - \bar{z}}{i(1 + |z|^2)}, \frac{|z|^2 - 1}{|z|^2 + 1}\right)$$

Prove that f is a homeomorphism.

Proof: We first show that f maps $\mathbb{C}$ into $\mathbb{S}^2$. We have

$$\left(\frac{z+\bar{z}}{1+|z|^2}\right)^2 + \left(\frac{z-\bar{z}}{i(1+|z|^2)}\right)^2 + \left(\frac{|z|^2-1}{|z|^2+1}\right)^2$$

$$= \frac{z^2 + 2z\bar{z} + \bar{z}^2}{1 + 2|z|^2 + |z|^4} + \frac{z^2 - 2z\bar{z} + \bar{z}^2}{-(1 + 2|z|^2 + |z|^4)} + \frac{1 - 2|z|^2 + |z|^4}{1 + 2|z|^2 + |z|^4}$$

$$= \frac{z^2 + 2z\bar{z} + \bar{z}^2}{1 + 2|z|^2 + |z|^4} + \frac{-z^2 + 2z\bar{z} - \bar{z}^2}{1 + 2|z|^2 + |z|^4} + \frac{1 - 2|z|^2 + |z|^4}{1 + 2|z|^2 + |z|^4}$$

$$= \frac{4z\bar{z} + 1 - 2|z|^2 + |z|^4}{1 + 2|z|^2 + |z|^4} = \frac{4|z|^2 + 1 - 2|z|^2 + |z|^4}{1 + 2|z|^2 + |z|^4} = \frac{1 + 2|z|^2 + |z|^4}{1 + 2|z|^2 + |z|^4} = 1.$$

So, f does in fact map $\mathbb{C}$ into $\mathbb{S}^2$.

To see that $N = (0, 0, 1) \notin \operatorname{ran} f$, note that if $f(z) = (0, 0, 1)$, then $z + \bar{z} = 0$ and $z - \bar{z} = 0$. Adding these two equations gives us that $2z = 0$, and so, $z = 0$. It follows that $\frac{|z|^2 - 1}{|z|^2 + 1} = -\frac{1}{1} = -1 \neq 1$.

We now show that f is a bijection by producing an inverse function. We define $g: \mathbb{S}^2 \setminus N \to \mathbb{C}$ by

$$g(a, b, c) = \left(\frac{a}{1-c}\right) + \left(\frac{b}{1-c}\right)i.$$

Observe that $\left(\frac{a}{1-c}\right) + \left(\frac{b}{1-c}\right)i \in \mathbb{C}$ unless $c = 1$, but if $c = 1$, then $a^2 + b^2 + c^2 = 1$ implies that $a^2 + b^2 = 0$, so that $a = b = 0$. But $N = (0, 0, 1)$ has been excluded from the domain of g.

Now, if $z = \left(\frac{a}{1-c}\right) + \left(\frac{b}{1-c}\right)i$, then $\bar{z} = \left(\frac{a}{1-c}\right) - \left(\frac{b}{1-c}\right)i$, so that $z + \bar{z} = \frac{2a}{1-c}$, $z - \bar{z} = \left(\frac{2b}{1-c}\right)i$, and $|z|^2 = \left(\frac{a}{1-c}\right)^2 + \left(\frac{b}{1-c}\right)^2 = \frac{a^2 + b^2}{(1-c)^2}$.

Assuming $a^2 + b^2 + c^2 = 1$, it follows that $a^2 + b^2 = 1 - c^2$. Therefore, we have

$$|z|^2 + 1 = \frac{a^2 + b^2}{(1-c)^2} + \frac{1 - 2c + c^2}{(1-c)^2} = \frac{1 - c^2}{(1-c)^2} + \frac{1 - 2c + c^2}{(1-c)^2} = \frac{2 - 2c}{(1-c)^2} = \frac{2(1-c)}{(1-c)^2} = \frac{2}{1-c}$$

and

$$|z|^2 - 1 = \frac{a^2 + b^2}{(1-c)^2} - \frac{1 - 2c + c^2}{(1-c)^2} = \frac{a^2 + b^2 - 1 + 2c - c^2}{(1-c)^2}$$

$$= \frac{1 - c^2 - 1 + 2c - c^2}{(1-c)^2} = \frac{2c - 2c^2}{(1-c)^2} = \frac{2c(1-c)}{(1-c)^2} = \frac{2c}{1-c}.$$

Thus, $\frac{z+\bar{z}}{1+|z|^2} = \frac{2a}{1-c} \cdot \frac{1-c}{2} = a$, $\frac{z-\bar{z}}{i(1+|z|^2)} = \left(\frac{2b}{1-c}\right)i \cdot \frac{(1-c)}{2i} = b$, and $\frac{|z|^2-1}{|z|^2+1} = \frac{2c}{1-c} \cdot \frac{1-c}{2} = c$.

Therefore, $f\big(g((a,b,c))\big) = f\left(\left(\frac{a}{1-c}\right) + \left(\frac{b}{1-c}\right)i\right) = \left(\frac{z+\bar{z}}{1+|z|^2}, \frac{z-\bar{z}}{i(1+|z|^2)}, \frac{|z|^2-1}{|z|^2+1}\right) = (a,b,c)$.

It follows that $f \circ g = id_{\mathbb{S}^2 \setminus N}$.

Now, if $(a,b,c) = \left(\frac{z+\bar{z}}{1+|z|^2}, \frac{z-\bar{z}}{i(1+|z|^2)}, \frac{|z|^2-1}{|z|^2+1}\right)$, then $1-c = 1 - \frac{|z|^2-1}{|z|^2+1} = \frac{|z|^2+1}{|z|^2+1} - \frac{|z|^2-1}{|z|^2+1} = \frac{2}{|z|^2+1}$.
Therefore, $\frac{a}{1-c} = \frac{z+\bar{z}}{1+|z|^2} \cdot \frac{|z|^2+1}{2} = \frac{z+\bar{z}}{2} = \operatorname{Re} z$ and $\frac{b}{1-c} = \frac{z-\bar{z}}{i(1+|z|^2)} \cdot \frac{|z|^2+1}{2} = \frac{z-\bar{z}}{2i} = \operatorname{Im} z$.

So, $g(f(z)) = g\left(\left(\frac{z+\bar{z}}{1+|z|^2}, \frac{z-\bar{z}}{i(1+|z|^2)}, \frac{|z|^2-1}{|z|^2+1}\right)\right) = \left(\frac{a}{1-c}\right) + \left(\frac{b}{1-c}\right) i = \operatorname{Re} z + (\operatorname{Im} z) i = z$.

It follows that $g \circ f = id_\mathbb{C}$.

Since $f \circ g = id_{\mathbb{S}^2 \setminus N}$ and $g \circ f = id_\mathbb{C}$, $g = f^{-1}$, and therefore, f is a bijection.

To see that f is continuous, first observe that $f = k \circ h$, where $h: \mathbb{C} \to \mathbb{C}^3$ is defined by $h(z) = (z,z,z)$ and $k: \mathbb{C}^3 \to \mathbb{R}^3$ is defined by $k(z,w,v) = \left(\frac{z+\bar{z}}{1+|z|^2}, \frac{w-\bar{w}}{i(1+|w|^2)}, \frac{|v|^2-1}{|v|^2+1}\right)$. We can verify this with the following computation: $(k \circ h)(z) = k(h(z)) = k(z,z,z) = \left(\frac{z+\bar{z}}{1+|z|^2}, \frac{z-\bar{z}}{i(1+|z|^2)}, \frac{|z|^2-1}{|z|^2+1}\right) = f(z)$.

Next note that in general, if X, Y, and Z are topological spaces with $j: X \to Y$ and $t: Y \to Z$ continuous functions. Then $t \circ j: X \to Z$ is continuous. To see this, let U be open in Z. Since t is continuous, we have $t^{-1}[U]$ open in Y. Since j is continuous, $j^{-1}[t^{-1}[U]]$ is open in X. Now, $x \in (t \circ j)^{-1}[U]$ if and only if $t(j(x)) = (t \circ j)(x) \in U$ if and only if $j(x) \in t^{-1}[U]$ if and only if $x \in j^{-1}[t^{-1}[U]]$. It follows that $(t \circ j)^{-1}[U] = j^{-1}[t^{-1}[U]]$. So, $(t \circ j)^{-1}[U]$ is open in X. Since U was an arbitrary open set in Z, we have shown that $t \circ j$ is continuous.

By the last paragraph, to see that f is continuous, it suffices to show that k and h are continuous.

To see that h is continuous, let U, V, and W be open in $\mathbb{C}$. Then $z \in h^{-1}(U \times V \times W)$ if and only if $(z,z,z) = h(z) \in U \times V \times W$ if and only if $z \in U$ and $z \in V$ and $z \in W$ if and only if $z \in U \cap V \cap W$. So, $h^{-1}(U \times V \times W) = U \cap V \cap W$, which is a finite intersection of open sets in $\mathbb{C}$, thus open. Since $\{A \times B \times C \mid A, B, C \text{ are open in } \mathbb{C}\}$ forms a basis for the product topology on $\mathbb{C}^3$, we have shown that h is continuous.

Next note that if X and Y are topological spaces with $j, t, s: X \to Y$ continuous functions, then the function $F: X^3 \to Y^3$ defined by $F(a,b,c) = (j(a), t(b), s(c))$ is continuous. To see this, let U, V, and W be arbitrary open sets in Y. Then we have $(a,b,c) \in F^{-1}(U \times V \times W)$ if and only if $(j(a), t(b), s(c)) = F((a,b,c)) \in U \times V \times W$ if and only if $j(a) \in U$, $t(b) \in V$ and $s(c) \in W$ if and only if $a \in j^{-1}[U]$ and $b \in t^{-1}[V]$ and $c \in s^{-1}[W]$ if and only if $(a,b,c) \in j^{-1}[U] \times t^{-1}[V] \times s^{-1}[W]$. So, $F^{-1}(U \times V \times W) = j^{-1}[U] \times t^{-1}[V] \times s^{-1}[W]$, which is open in the product topology of X^3. Since $\{A \times B \times C \mid A, B, C \text{ are open in } Y\}$ forms a basis for the product topology on Y^3, we have shown that F is continuous.

By the last paragraph, to see that k is continuous, it suffices to show that the functions $j, t, s: \mathbb{C} \to \mathbb{R}$ defined by $j(z) = \frac{z+\bar{z}}{1+|z|^2}$, $t(z) = \frac{z-\bar{z}}{i(1+|z|^2)}$, and $s(z) = \frac{|z|^2-1}{|z|^2+1}$ are continuous.

Let's prove that j is continuous directly from the definition of continuity. Let $w = a + bi \in \mathbb{C}$, let $\epsilon > 0$, let $M = \max\{1, |2a-1|, |2a+1|, |2b-1|, |2b+1|, 1+a^2+b^2\}$, and let $\delta = \min\left\{1, \frac{\epsilon}{6M^2}\right\}$. Suppose that $|z - w| < \delta$. It follows that $|(x + yi) - (a + bi)| < \delta$, or equivalently, $|(x-a) + (y-b)i| < \delta$. So, $(x-a)^2 + (y-b)^2 < \delta^2$. Since $|x-a|$ and $|y-b|$ are both less than or equal to $\sqrt{(x-a)^2 + (y-b)^2} < \delta$, we have $|x - a| < \delta$ and $|y - b| < \delta$. Since $\delta \le 1$, $|x - a| < 1$, or equivalently, $-1 < x - a < 1$. Adding $2a$, we have $2a - 1 < x + a < 2a + 1$. So, $|x + a| < \max\{|2a-1|, |2a+1|\} \le M$. Similarly, $|y + b| < \max\{|2b-1|, |2b+1|\} \le M$. We also have $|a| < a^2 + b^2 + 1 \le M$. Also, since $M \ge 1$, $M^2 \ge M$. So, we have

$$\left| \frac{z + \bar{z}}{1 + |z|^2} - \frac{w + \bar{w}}{1 + |w|^2} \right| = \left| \frac{2x}{1 + x^2 + y^2} - \frac{2a}{1 + a^2 + b^2} \right| = \left| \frac{2x(1 + a^2 + b^2) - 2a(1 + x^2 + y^2)}{(1 + x^2 + y^2)(1 + a^2 + b^2)} \right|$$
$$\le |2x(1 + a^2 + b^2) - 2a(1 + x^2 + y^2)|$$
$$= |(2x - 2a)(1 + a^2 + b^2) + 2a(1 + a^2 + b^2) - 2a(1 + x^2 + y^2)|$$
$$= |2(x - a)(1 + a^2 + b^2) + 2a(a^2 + b^2 - x^2 - y^2)|$$
$$\le 2|x - a||1 + a^2 + b^2| + 2|a||a - x||a + x| + 2|a||b - y||b + y|$$
$$< 2 \cdot \frac{\epsilon}{6M^2} \cdot M^2 + 2M \cdot \frac{\epsilon}{6M^2} \cdot M + 2M \cdot \frac{\epsilon}{6M^2} \cdot M = \epsilon.$$

So, j is continuous at $w = a + bi \in \mathbb{C}$. Since $w \in \mathbb{C}$ was arbitrary, j is continuous on $\mathbb{C}$.

The proofs that t and s are continuous are similar. It follows that f is continuous.

To see that $g = f^{-1}$ is continuous, first observe that $g(a, b, c) = u(a, b, c) + iv(a, b, c)$, where $u: \mathbb{S}^2 \setminus N \to \mathbb{R}$ and $v: \mathbb{S}^2 \setminus N \to \mathbb{R}$ are defined by $u(a, b, c) = \frac{a}{1-c}$ and $v(a, b, c) = \frac{b}{1-c}$. The proof used in Problem 17 can be modified slightly to show that g is continuous if and only if u and v are continuous.

We leave it to the reader to show that u and v are continuous. It then follows that $g = f^{-1}$ is continuous.

Since $f: \mathbb{C} \to \mathbb{S}^2 \setminus N$ is bijective, continuous and has a continuous inverse, f is a homeomorphism. $\square$

Problem Set 16

LEVEL 1

1. Let V and W be vector spaces over $\mathbb{R}$. Determine if each of the following functions is a linear transformation: (i) $f: \mathbb{R} \to \mathbb{R}$ defined by $f(x) = 2x + 1$; (ii) $g: \mathbb{R} \to \mathbb{R}^2$ defined by $g(x) = (2x, 3x)$; (iii) $h: \mathbb{R}^3 \to \mathbb{R}^3$ defined by $h((x, y, z)) = (x + y, x + z, z - y)$.

Solutions:

(i) $f(0) = 2 \cdot 0 + 1 = 1 \neq 0$. Since the image of 0 under a linear transformation is 0, f is **not** a linear transformation.

(ii) $g(ax + by) = (2(ax + by), 3(ax + by)) = (2ax + 2by, 3ax + 3by)$
$= (2ax, 3ax) + (2by, 3by) = a(2x, 3x) + b(2y, 3y) = ag(x) + bg(y)$.

By Theorem 16.1, g is a linear transformation.

(iii) $h(a(x, y, z) + b(s, t, w)) = h((ax + bs, ay + bt, az + bw))$
$= (ax + bs + ay + bt, ax + bs + az + bw, az + bw - ay - bt)$
$= (ax + ay, ax + az, az - ay) + (bs + bt, bs + bw, bw - bt)$
$a(x + y, x + z, z - y) + b(s + t, s + w, w - t) = ah((x, y, z)) + bh((s, t, w))$.

By Theorem 16.1, h is a linear transformation.

2. Compute each of the following: (i) $\begin{bmatrix} 2 & 0 & -3 \\ 0 & 1 & 4 \end{bmatrix} \cdot \begin{bmatrix} 1 & 1 & 3 & 0 \\ 1 & -4 & 2 & 0 \\ 2 & 0 & 1 & -4 \end{bmatrix}$; (ii) $\begin{bmatrix} 3 & -1 & 5 \end{bmatrix} \cdot \begin{bmatrix} -4 \\ -7 \\ 2 \end{bmatrix}$;

(iii) $\begin{bmatrix} -4 \\ -7 \\ 2 \end{bmatrix} \cdot \begin{bmatrix} 3 & -1 & 5 \end{bmatrix}$; (iv) $\begin{bmatrix} a & b & c \\ d & e & f \\ g & h & i \end{bmatrix} \cdot \begin{bmatrix} 1 & 0 & 1 \\ 0 & 2 & 0 \\ 3 & 1 & 4 \end{bmatrix}$.

Solutions:

(i) $\begin{bmatrix} 2 & 0 & -3 \\ 0 & 1 & 4 \end{bmatrix} \cdot \begin{bmatrix} 1 & 1 & 3 & 0 \\ 1 & -4 & 2 & 0 \\ 2 & 0 & 1 & -4 \end{bmatrix} = \begin{bmatrix} -4 & 2 & 3 & 12 \\ 9 & -4 & 6 & -16 \end{bmatrix}$.

(ii) $\begin{bmatrix} 3 & -1 & 5 \end{bmatrix} \cdot \begin{bmatrix} -4 \\ -7 \\ 2 \end{bmatrix} = -12 + 7 + 10 = 5$.

(iii) $\begin{bmatrix} -4 \\ -7 \\ 2 \end{bmatrix} \cdot \begin{bmatrix} 3 & -1 & 5 \end{bmatrix} = \begin{bmatrix} -12 & 4 & -20 \\ -21 & 7 & -35 \\ 6 & -2 & 10 \end{bmatrix}$.

(iv) $\begin{bmatrix} a & b & c \\ d & e & f \\ g & h & i \end{bmatrix} \cdot \begin{bmatrix} 1 & 0 & 1 \\ 0 & 2 & 0 \\ 3 & 1 & 4 \end{bmatrix} = \begin{bmatrix} a + 3c & 2b + c & a + 4c \\ d + 3f & 2e + f & d + 4f \\ g + 3i & 2h + i & g + 4i \end{bmatrix}$.

LEVEL 2

3. Consider $\mathbb{C}$ as a vector space over itself. Give an example of a function $f: \mathbb{C} \to \mathbb{C}$ such that f is additive, but not a linear transformation. Then give an example of vector spaces V and W and a homogenous function $g: V \to W$ that is **not** a linear transformation.

Solution: Define $f: \mathbb{C} \to \mathbb{C}$ by $f(z) = \text{Re } z$. If $z = x + yi$ and $w = u + vi$, then

$$f(z+w) = f\big((x+yi) + (u+vi)\big) = f\big((x+u) + (y+v)i\big)$$
$$= x + u = \text{Re } z + \text{Re } w = f(z) + f(w).$$

So, f is additive.

Now, $f(i \cdot 1) = f(i) = 0$ and $i \cdot f(1) = i \cdot 1 = i$. So, $f(i \cdot 1) \neq i \cdot f(1)$. Therefore, f is **not** homogenous.

So, f is **not** a linear transformation.

Let $V = \mathbb{R}^2$ and $W = \mathbb{R}$, considered as vector spaces over the field $\mathbb{R}$. Define $g: \mathbb{R}^2 \to \mathbb{R}$ by $g\big((v,w)\big) = (v^3 + w^3)^{\frac{1}{3}}$. Then for $k \in \mathbb{R}$, we have

$$g\big(k(v,w)\big) = g\big((kv, kw)\big) = ((kv)^3 + (kw)^3)^{\frac{1}{3}} = (k^3 v^3 + k^3 w^3)^{\frac{1}{3}} = \big(k^3(v^3 + w^3)\big)^{\frac{1}{3}}$$
$$= (k^3)^{\frac{1}{3}}(v^3 + w^3)^{\frac{1}{3}} = k(v^3 + w^3)^{\frac{1}{3}} = k \cdot g\big((v,w)\big).$$

Therefore, g is homogeneous.

Now, we have

$$g\big((1,0)\big) + g\big((0,1)\big) = (1^3 + 0^3)^{\frac{1}{3}} + (0^3 + 1^3)^{\frac{1}{3}} = 1 + 1 = 2.$$
$$g\big((1,0) + (0,1)\big) = g\big((1,1)\big) = (1^3 + 1^3)^{\frac{1}{3}} = 2^{\frac{1}{3}}.$$

So, $g\big((1,0) + (0,1)\big) \neq g\big((1,0)\big) + g\big((0,1)\big)$, and therefore, g is **not** additive.

So, g is **not** a linear transformation.

LEVEL 3

4. Let $P = \{ax^2 + bx + c \mid a, b, c \in \mathbb{R}\}$ be the vector space of polynomials of degree at most 2 with real coefficients (see part 3 of Example 8.3 from Lesson 8). Define the linear transformation $D: P \to P$ by $D(ax^2 + bx + c) = 2ax + b$. Find the matrix of T with respect to each of the following bases: (i) The standard basis $B = \{1, x, x^2\}$; (ii) $C = \{x+1, x^2+1, x^2+x\}$.

143

Solutions:

(i) $D(1) = 0 = 0x^2 + 0x + 0$, $D(x) = 1 = 0x^2 + 0x + 1$, and $D(x^2) = 2x = 0x^2 + 2x + 0$.

Therefore, $\mathcal{M}_T = \begin{bmatrix} 0 & 0 & 0 \\ 0 & 0 & 2 \\ 0 & 1 & 0 \end{bmatrix}$.

(ii) We have the following:
$$D(x+1) = 1 = \frac{1}{2}(x+1) + \frac{1}{2}(x^2+1) - \frac{1}{2}(x^2+x)$$
$$D(x^2+1) = 2x = 1(x+1) - 1(x^2+1) + 1(x^2+x)$$
$$D(x^2+x) = 2x+1 = \frac{3}{2}(x+1) - \frac{1}{2}(x^2+1) + \frac{1}{2}(x^2+x)$$

Therefore, $\mathcal{M}_T(C) = \begin{bmatrix} \frac{1}{2} & 1 & \frac{3}{2} \\ \frac{1}{2} & -1 & -\frac{1}{2} \\ -\frac{1}{2} & 1 & \frac{1}{2} \end{bmatrix}$.

Notes: (1) In part (ii), it's a little more challenging to express the images of the given basis elements as a linear combination of the given basis elements. For example, how do we express 1 as a linear combination of $x+1$, x^2+1, and x^2+x. Well, we need to find weights c_1, c_2, and c_3 such that

$$1 = c_1(x+1) + c_2(x^2+1) + c_3(x^2+x).$$

We rewrite each side of this equation as a linear combination of x^2, x, and 1 as follows:

$$1 = c_1 x + c_1 + c_2 x^2 + c_2 + c_3 x^2 + c_3 x$$
$$0x^2 + 0x + 1 = (c_2 + c_3)x^2 + (c_1 + c_3)x + (c_1 + c_2)$$

Since $x^2, x, 1$ are linearly independent, we must have $c_2 + c_3 = 0$, $c_1 + c_3 = 0$, and $c_1 + c_2 = 1$. Subtracting the first equation from the third equation gives $c_1 - c_3 = 1$. Adding this equation to the second equation gives $2c_1 = 1$. Dividing by 2, we get $c_1 = \frac{1}{2}$. Using this value for c_1 and the second and third equations, we get $c_2 = \frac{1}{2}$ and $c_3 = -\frac{1}{2}$. This is how we wrote $D(x+1)$ as a linear combination of $x+1$, x^2+1, and x^2+x in the solution above.

The computations for $D(x^2+1)$ and $D(x^2+x)$ can be done similarly.

(2) In a computational linear algebra course, you learn a procedure for solving systems of linear equations called Gauss-Jordan reduction. This procedure will always allow you to express arbitrary vectors as linear combinations of basis vectors. In fact, a single Gauss-Jordan reduction can be used to express the images of as many vectors as we like as a linear combination of basis vectors. Performing tedious computations like this lies outside the scope of this book and so, we leave it for the interested reader to investigate themselves.

5. Let V and W be vector spaces with V finite-dimensional, let $U \leq V$, and let $T \in \mathcal{L}(U, W)$. Prove that there is an $S \in \mathcal{L}(V, W)$ such that $S(v) = T(v)$ for all $v \in U$.

Proof: Suppose that $U \leq V$ with $\dim V = n$. Let $B = \{v_1, v_2, \ldots, v_k\}$ be a basis of U. Since $v_1, v_2, \ldots, v_k$ are linearly independent, by the note following Theorem 16.8, we can extend B to a basis B' of V, say $B' = \{v_1, v_2, \ldots, v_k, v_{k+1}, \ldots, v_n\}$. Define S by $S(c_1 v_1 + \cdots + c_n v_n) = T(c_1 v_1 + \cdots + c_k v_k)$. Since every vector in V can be written as a linear combination of the vectors in B', $\dom S = V$. Also, $\ran S \subseteq \ran T \subseteq W$. So, $S: V \to W$. Since T is linear, so is S. If $v \in U$, then v can be written as a linear combination of the vectors in B, say $v = c_1 v_1 + \cdots c_k v_k$. Then we have

$$S(v) = S(c_1 v_1 + \cdots + c_k v_k) = S(c_1 v_1 + \cdots + c_k v_k + 0 v_{k+1} + \cdots + 0 v_n)$$
$$= T(c_1 v_1 + \cdots + c_k v_k) = T(v). \qquad \square$$

LEVEL 4

6. Let $T: V \to W$ be a linear transformation and let $v_1, v_2, \ldots, v_n \in V$. Prove the following: (i) If T is injective and $v_1, v_2, \ldots, v_n$ are linearly independent in V, then $T(v_1), T(v_2), \ldots, T(v_n)$ are linearly independent in W. (ii) If T is surjective and $\span\{v_1, v_2, \ldots, v_n\} = V$, then $\span\{T(v_1), T(v_2), \ldots, T(v_n)\} = W$.

Proofs:

(i) Let T be injective and assume that $v_1, v_2, \ldots, v_n$ are linearly independent. Let $c_1, \ldots, c_n \in \mathbb{F}$ be such that $c_1 T(v_1) + \cdots + c_n T(v_n) = 0$. Since T is a linear transformation, $T(c_1 v_1 + \cdots + c_n v_n) = 0$. Since T is injective and $T(0) = 0$, we have $c_1 v_1 + \cdots + c_n v_n = 0$. By the linear independence of $v_1, v_2, \ldots, v_n$, $c_1 = c_2 = \cdots = c_n = 0$. So, $T(v_1), T(v_2), \ldots, T(v_n)$ are linearly independent. $\qquad \square$

(ii) Let T be surjective and assume that $\span\{v_1, v_2, \ldots, v_n\} = V$. Let $w \in W$. Since T is surjective, there is $v \in V$ such that $T(v) = w$. Since $\span\{v_1, v_2, \ldots, v_n\} = V$, $v = c_1 v_1 + \cdots + c_n v_n$ for some $c_1, \ldots, c_n \in \mathbb{F}$. Then $T(v) = T(c_1 v_1 + \cdots + c_n v_n) = c_1 T(v_1) + \cdots + c_n T(v_n)$. Therefore, $w = T(v)$ is in $\span\{T(v_1), T(v_2), \ldots, T(v_n)\}$. So, $W \subseteq \span\{T(v_1), T(v_2), \ldots, T(v_n)\}$. Since $T: V \to W$ and W is closed under taking linear combinations, $\span\{T(v_1), T(v_2), \ldots, T(v_n)\} \subseteq W$. Thus, we have $\span\{T(v_1), T(v_2), \ldots, T(v_n)\} = W$. $\qquad \square$

7. Determine if each linear transformation is diagonalizable: (i) $T: \mathbb{R}^2 \to \mathbb{R}^2$ defined by $T((x, y)) = (y, 2x)$; (ii) $U: \mathbb{C}^2 \to \mathbb{C}^2$ defined by $U((z, w)) = (z + iw, iz - w)$.

Proofs:

(i) We find the eigenvalues of T by solving the equation $T((x, y)) = \lambda(x, y)$, or equivalently, $(y, 2x) = (\lambda x, \lambda y)$. Equating the first components and second components gives us the two equations $y = \lambda x$ and $2x = \lambda y$. Substituting λx for y in the second equation gives $2x = \lambda^2 x$, or equivalently, $\lambda^2 x - 2x = 0$. Using distributivity on the left-hand side of this equation gives $x(\lambda^2 - 2) = 0$. So, $x = 0$ or $\lambda^2 - 2 = 0$. If $x = 0$, then $y = \lambda(0) = 0$. So, $(x, y) = (0, 0)$. Since an eigenvector must be nonzero, we reject $x = 0$. The equation $\lambda^2 - 2 = 0$ has the two solutions $\lambda = \sqrt{2}$ and $\lambda = -\sqrt{2}$. These are the two eigenvalues of T.

Next, let's find the eigenvector corresponding to the eigenvalue $\lambda = \sqrt{2}$. In this case, we have $T((x,y)) = \sqrt{2}(x,y)$, or equivalently, $(y, 2x) = (\sqrt{2}x, \sqrt{2}y)$. So, $y = \sqrt{2}x$ and $2x = \sqrt{2}y$. These two equations are actually equivalent. Indeed, if we multiply each side of the first equation by $\sqrt{2}$, we get $\sqrt{2}y = \sqrt{2} \cdot \sqrt{2}x$, or equivalently, $\sqrt{2}y = 2x$.

So, we use only one of the equations, say $y = \sqrt{2}x$. So, the eigenvectors of T corresponding to the eigenvalue $\lambda = \sqrt{2}$ are all nonzero vectors of the form $(x, \sqrt{2}x)$. In particular, if we let $x = 1$, we get the eigenvector $(1, \sqrt{2})$.

Let's also find the eigenvector corresponding to the eigenvalue $\lambda = -\sqrt{2}$. In this case, we have $T((x, y)) = -\sqrt{2}(x, y)$, or equivalently, $(y, 2x) = (-\sqrt{2}x, -\sqrt{2}y)$. So, $y = -\sqrt{2}x$ and $2x = -\sqrt{2}y$. These two equations are equivalent. Indeed, if we multiply each side of the first equation by $-\sqrt{2}$, we get $-\sqrt{2}y = -\sqrt{2}(-\sqrt{2}x)$, or equivalently, $-\sqrt{2}y = 2x$.

So, we use only one of the equations, say $y = -\sqrt{2}x$. So, the eigenvectors of T corresponding to the eigenvalue $\lambda = -\sqrt{2}$ are all nonzero vectors of the form $(x, -\sqrt{2}x)$. In particular, if we let $x = 1$, we get the eigenvector $(1, -\sqrt{2})$.

It follows that $B = \{(1, \sqrt{2}), (1, -\sqrt{2})\}$ is a basis of eigenvectors of V and we have

$$T((1, \sqrt{2})) = \sqrt{2}(1, \sqrt{2})$$
$$T((1, -\sqrt{2})) = -\sqrt{2}(1, -\sqrt{2})$$

Therefore, the matrix of T with respect to B is $\mathcal{M}_T(B) = \begin{bmatrix} \sqrt{2} & 0 \\ 0 & -\sqrt{2} \end{bmatrix}$.

Since $\mathcal{M}_T(B)$ is a diagonal matrix, T is diagonalizable.

(ii) We find the eigenvalues of U by solving the equation $U((z, w)) = \lambda(z, w)$, or equivalently, $(z + iw, iz - w) = (\lambda z, \lambda w)$. Equating the first components and second components gives us the two equations $z + iw = \lambda z$ and $iz - w = \lambda w$. From the second equation, we must have $\lambda w + w = iz$, or equivalently, $w(\lambda + 1) = iz$. Multiplying the first equation by i gives the equation $iz - w = \lambda iz$. Replacing iz by $w(\lambda + 1)$ gives us $w(\lambda + 1) - w = \lambda w(\lambda + 1)$, or equivalently, $w(\lambda + 1 - 1) = w\lambda(\lambda + 1)$. So, we get $w\lambda - w\lambda(\lambda + 1) = 0$, or equivalently, $w\lambda(1 - \lambda - 1) = 0$ or $-w\lambda^2 = 0$. So, $\lambda = 0$ or $w = 0$. Since $w = 0$ implies $z = 0$ and an eigenvector must be nonzero, we reject $w = 0$. It follows that $\lambda = 0$ is the only eigenvalue of U.

Let's find the eigenvectors corresponding to the eigenvalue $\lambda = 0$. In this case, we have $U((z, w)) = 0(z, w)$, or equivalently, $(z + iw, iz - w) = (0, 0)$. Therefore, $z + iw = 0$ and $iz - w = 0$. These two equations are actually equivalent. Indeed, if we multiply each side of the first equation by i, we get $iz - w = 0$.

So, we use only one of the equations, say $iz - w = 0$, or equivalently, $w = iz$ So, the eigenvectors of U corresponding to the eigenvalue $\lambda = 0$ are all nonzero vectors of the form $(z, iz) = z(1, i)$. Since all these vectors are scalar multiples of each other, a basis of eigenvectors has just one vector. Therefore, U is **not** diagonalizable.

8. Let V and W be vector spaces over a field $\mathbb{F}$. Prove that $\mathcal{L}(V, W)$ is a vector space over $\mathbb{F}$, where addition and scalar multiplication are defined as in Theorem 16.2.

Proof: We first prove that $(\mathcal{L}(V, W), +)$ is a commutative group.

(Closure) Let $S, T \in \mathcal{L}(V, W)$, let $v, w \in V$, and let $a, b \in \mathbb{F}$. Then
$$(S + T)(av + bw) = S(av + bw) + T(av + bw) = aS(v) + bS(w) + aT(v) + bT(w)$$
$$= a\big(S(v) + T(v)\big) + b\big(S(w) + T(w)\big) = a(S + T)(v) + b(S + T)(w).$$

So, $S + T \in \mathcal{L}(V, W)$.

(Associativity) Let $S, T, U \in \mathcal{L}(V, W)$ and let $v \in V$. Since addition is associative in W, we have
$$\big((S + T) + U\big)(v) = (S + T)(v) + U(v) = \big(S(v) + T(v)\big) + U(v)$$
$$= S(v) + \big(T(v) + U(v)\big) = S(v) + (T + U)(v) = \big(S + (T + U)\big)(v).$$

So, $(S + T) + U = S + (T + U)$.

(Commutativity) Let $S, T \in \mathcal{L}(V, W)$ and let $v \in V$. Since addition is commutative in W, we have
$$(S + T)(v) = S(v) + T(v) = T(v) + S(v) = (T + S)(v).$$

So, $S + T = T + S$.

(Identity) Define $0 \colon V \to W$ by $0(v) = 0$ for all $v \in V$. Then for all $v, w \in V$ and $a, b \in \mathbb{F}$,
$$0(av + bw) = 0 = 0 + 0 = a \cdot 0 + b \cdot 0 = a0(v) + b0(w).$$

So, $0 \in \mathcal{L}(V, W)$.

For any $T \in \mathcal{L}(V, W)$ and $v \in V$, we have $(T + 0)(v) = T(v) + 0(v) = T(v) + 0 = T(v)$ and we have $(0 + T)(v) = 0(v) + T(v) = 0 + T(v) = T(v)$. So, $T + 0 = 0 + T = T$.

(Inverse) Let $T \in \mathcal{L}(V, W)$ and define S by $S(v) = -T(v)$ for all $v \in V$. Then for all $v, w \in V$ and $a, b \in \mathbb{F}$,
$$S(av + bw) = -T(av + bw) = -\big(aT(v) + bT(w)\big) = a\big(-T(v)\big) + b\big(-T(w)\big) = aS(v) + bS(w).$$

So, $S \in \mathcal{L}(V, W)$. If $v \in V$, then
$$(T + S)(v) = T(v) + S(v) = T(v) + \big(-T(v)\big) = 0 = 0(v).$$
$$(S + T)(v) = S(v) + T(v) = -T(v) + T(v) = 0 = 0(v).$$

So, $T + S = S + T = 0$. Therefore, $S = -T$.

Now, let's prove that $\mathcal{L}(V, W)$ has the remaining vector space properties.

(Closure under scalar multiplication) Let $k \in \mathbb{F}$, let $T \in \mathcal{L}(V,W)$, let $v, w \in V$, and let $a, b \in \mathbb{F}$. Then
$$(kT)(av + bw) = kT(av + bw) = k\big(aT(v) + bT(w)\big) = k\big(aT(v)\big) + k\big(bT(w)\big)$$
$$= (ka)T(v) + (kb)T(w) = (ak)T(v) + (bk)\big(T(w)\big) = a\big(kT(v)\big) + b\big(kT(w)\big)$$
$$= a(kT)(v) + b(kT)(w).$$
So, $kT \in \mathcal{L}(V,W)$.

(Scalar multiplication identity) Let 1 be the multiplicative identity of $\mathbb{F}$, let $T \in \mathcal{L}(V,W)$, and let $v \in V$. Then $(1T)(v) = 1T(v) = T(v)$. So, $1T = T$.

(Associativity of scalar multiplication) Let $j, k \in \mathbb{F}$, let $T \in \mathcal{L}(V,W)$, and let $v \in V$. Then since multiplication is associative in W, we have
$$\big((jk)T\big)(v) = (jk)T(v) = j\big(kT(v)\big) = j(kT)(v) = \big(j(kT)\big)(v).$$
So, $(jk)T = j(kT)$.

(Distributivity of 1 scalar over 2 vectors) Let $k \in \mathbb{F}$, let $S, T \in \mathcal{L}(V,W)$, and let $v \in V$. Since multiplication is distributive over addition in W, we have
$$\big(k(S+T)\big)(v) = k(S+T)(v) = k\big(S(v) + T(v)\big) = kS(v) + kT(v)$$
$$= (kS)(v) + (kT)(v) = (kS + kT)(v).$$
So, $k(S + T) = kS + kT$.

(Distributivity of 2 scalars over 1 vector) Let $j, k \in \mathbb{F}$, let $S \in \mathcal{L}(V,W)$, and let $v \in V$. Since multiplication is distributive over addition in W, we have
$$\big((j+k)S\big)(v) = (j+k)S(v) = jS(v) + kS(v) = (jS)(v) + (kS)(v) = (jS + kS)(v).$$
So, $(j + k)S = jS + kS$. $\square$

9. Let V be a vector space over a field $\mathbb{F}$. Prove that $\mathcal{L}(V)$ is a linear algebra over $\mathbb{F}$, where addition and scalar multiplication are defined as in Theorem 16.2 and vector multiplication is given by composition of linear transformations.

Proof: By Problem 8, $(\mathcal{L}(V), +)$ is a vector space over $\mathbb{F}$. We now go through the properties for vector multiplication.

(Closure) We showed right before Example 16.2 that if V, W, and U are vector spaces over $\mathbb{F}$, and $T: V \to W$, $S: W \to U$ are linear transformations, then the composition $S \circ T: V \to W$ is a linear transformation. If $S, T \in \mathcal{L}(V)$, then $V = W = U$, and so, $S \circ T: V \to V$. Therefore, $ST = S \circ T \in \mathcal{L}(V)$.

(Associativity) Let $S, T, U \in \mathcal{L}(V)$ and let $v \in V$. Then
$$\big((ST)U\big)(v) = (ST)(U(v)) = S\big(T(U(v))\big) = S((TU)(v)) = \big(S(TU)\big)(v).$$
Since $v \in V$ was arbitrary, $(ST)U = S(TU)$.

(Identity) Define $I: V \to V$ by $I(v) = v$ for all $v \in V$. Then for all $v, w \in V$ and $a, b \in \mathbb{F}$,
$$I(av + bw) = av + bw = aI(v) + bI(w).$$
So, $I \in \mathcal{L}(V)$.

For any $T \in \mathcal{L}(V)$ and $v \in V$, we have $(TI)(v) = T(I(v)) = T(v)$ and $(IT)(v) = I(T(v)) = T(v)$. So, $TI = IT = T$.

(Left Distributivity) Let $S, T, U \in \mathcal{L}(V)$ and let $v \in V$. Since multiplication is distributive over addition in V, we have
$$\big(S(T + U)\big)(v) = S\big((T + U)(v)\big) = S\big(T(v) + U(v)\big) = S\big(T(v)\big) + S\big(U(v)\big)$$
$$= (ST)(v) + (SU)(v) = (ST + SU)(v).$$
Since $v \in V$ was arbitrary, $S(T + U) = ST + SU$.

(Right Distributivity) Let $S, T, U \in \mathcal{L}(V)$ and let $v \in V$. Since multiplication is distributive over addition in V, we have
$$\big((S + T)U\big)(v) = (S + T)\big(U(v)\big) = S\big(U(v)\big) + T\big(U(v)\big) = (SU)(v) + (TU)(v) = (SU + TU)(v).$$
Since $v \in V$ was arbitrary, $(S + T)U = SU + TU$.

(Compatibility of scalar and vector multiplication) Let $S, T \in \mathcal{L}(V)$ and $k \in \mathbb{F}$.
$$\big(k(ST)\big)(v) = k\big((ST)(v)\big) = k\big(S(T(v))\big) = (kS)\big(T(v)\big) = \big((kS)T\big)(v)$$
So, $k(ST) = (kS)T$.
$$\big(k(ST)\big)(v) = k\big((ST(v))\big) = k\big(S(T(v))\big) = S\big(k(T(v))\big) = S\big((kT)(v)\big) = \big(S(kT)\big)(v)$$
So, $k(ST) = S(kT)$. $\square$

10. Let $T: V \to W$ and $S: W \to V$ be linear transformations such that $ST = i_V$ and $TS = i_W$. Prove that S and T are bijections and that $S = T^{-1}$.

Proof: By symmetry, it suffices to show that T is a bijection.

Let $v, w \in V$ and suppose that $T(v) = T(w)$. Since $ST = i_V$,
$$v = i_V(v) = (ST)(v) = S\big(T(v)\big) = S\big(T(w)\big) = (ST)(w) = i_V(w) = w.$$
Therefore, T is injective.

Let $w \in W$. Since $TS = i_W$, $w = i_W(w) = (TS)(w) = T\big(S(w)\big)$. So, $w \in \operatorname{ran} T$. Therefore, T is surjective.

Since T is injective and surjective, T is a bijection.

Suppose that $T(v) = w$. Then $v = i_V(v) = (ST)(v) = S\big(T(v)\big) = S(w)$. So, $S = T^{-1}$. $\square$

11. Let V and W be finite-dimensional vector spaces and let $T \in \mathcal{L}(V, W)$. Prove the following: (i) If $\dim V < \dim W$, then T is **not** surjective. (ii) If $\dim V > \dim W$, then T is **not** injective.

Proofs:

(i) By Theorem 16.8, $\operatorname{rank} T + \operatorname{nullity} T = \dim V$. So, we have
$$\operatorname{rank} T = \dim V - \operatorname{nullity} T \leq \dim V < \dim W.$$
So, $\dim T[V] = \operatorname{rank} T < \dim W$. Therefore, $T[V] \neq W$, and so, T is **not** surjective. □

(ii) By Theorem 16.8, $\operatorname{rank} T + \operatorname{nullity} T = \dim V$. So, we have
$$\operatorname{nullity} T = \dim V - \operatorname{rank} T \geq \dim V - \dim W > 0.$$
So, $\dim \ker(T) > 0$. Therefore, $\ker(T) \neq \{0\}$. By Theorem 16.7, T is **not** injective. □

12. Prove that two finite-dimensional vector spaces over a field $\mathbb{F}$ are isomorphic if and only if they have the same dimension.

Proof: Let V and W be finite-dimensional vector spaces over a field $\mathbb{F}$. First suppose that $\dim V = \dim W = n$. By part 2 of Example 16.4, both V and W are isomorphic to $\mathbb{F}^n$. Since isomorphism is an equivalence relation, V and W are isomorphic to each other.

Conversely, suppose that V and W are isomorphic. Let $T: V \to W$ be an isomorphism. Since T is an isomorphism, it is surjective. By part (i) of Problem 11, $\dim V \geq \dim W$. Since T is an isomorphism, it is injective. By part (ii) of Problem 11, $\dim V \leq \dim W$. Therefore, $\dim V = \dim W$.

13. Let $T \in \mathcal{L}(V)$ be invertible and let $\lambda \in \mathbb{F} \setminus \{0\}$. Prove that λ is an eigenvalue of T if and only if $\frac{1}{\lambda}$ is an eigenvalue of T^{-1}.

Proof: λ is an eigenvalue of T if and only if there is a nonzero $v \in V$ such that $T(v) = \lambda v$. Now,
$$T(v) = \lambda v \Leftrightarrow T^{-1}(T(v)) = T^{-1}(\lambda v) \Leftrightarrow (T^{-1}T)(v) = \lambda T^{-1}(v)$$
$$\Leftrightarrow i_V(v) = \lambda T^{-1}(v) \Leftrightarrow v = \lambda T^{-1}(v) \Leftrightarrow \frac{1}{\lambda} v = T^{-1}(v).$$

So, λ is an eigenvalue of T if and only if $T^{-1}(v) = \frac{1}{\lambda} v$ if and only if $\frac{1}{\lambda}$ is an eigenvalue of T^{-1}. □

LEVEL 5

14. Let V be a vector space with $\dim V > 1$. Show that $\{T \in \mathcal{L}(V) \mid T \text{ is not invertible}\} \not\leq \mathcal{L}(V)$.

Proof: Let $X = \{T \in \mathcal{L}(V) \mid T \text{ is not invertible}\}$. First suppose that $\dim V = 2$ and let $\{v_1, v_2\}$ be a basis of V. Let T be the linear transformation such that $T(v_1) = v_1$ and $T(v_2) = v_1$. Since T is not injective, T is not invertible. Let U be the linear transformation such that $U(v_1) = 0$ and $U(v_2) = -v_1 + v_2$. Since $v_1 \neq 0$ and $U(0) = 0$, U is not injective, and therefore, U is not invertible. Now, we have $(T + U)(v_1) = T(v_1) + U(v_1) = v_1 + 0 = v_1$, $(T + U)(v_2) = T(v_2) + U(v_2) = v_1 - v_1 + v_2 = v_2$. So, $T + U = i_V$, which is invertible. Therefore, X is not closed under addition, and so, $X \not\leq \mathcal{L}(V)$.

Now, let $\dim V = n > 2$. Let T be the linear transformation such that $T(v_1) = T(v_2) = v_1$ and $T(v_i) = v_i$ for each $i = 3, \ldots, n$. Since T is not injective, T is not invertible. Let U be the linear transformation such that $U(v_2) = -v_1 + v_2$ and $U(v_i) = 0$ for each $i \neq 2$. Since U is not injective, U is not invertible. Now, we have $(T + U)(v_2) = T(v_2) + U(v_2) = v_1 - v_1 + v_2 = v_2$ and for all $i \neq 2$, $(T + U)(v_i) = T(v_i) + U(v_i) = v_i + 0 = v_i$. So, $T + U = i_V$, which is invertible. Therefore, X is not closed under addition, and so, $X \nsubseteq \mathcal{L}(V)$. □

15. Let V be an n-dimensional vector space over a field $\mathbb{F}$. Prove that there is a linear algebra isomorphism $F: \mathcal{L}(V) \to M_{nn}^{\mathbb{F}}$.

Proof: Let $B = \{v_1, \ldots, v_n\}$ be a basis of V and define $F: \mathcal{L}(V) \to M_{nn}^{\mathbb{F}}$ by $F(T) = \mathcal{M}_T(B)$.

Suppose that $F(T) = F(U)$, so that $\mathcal{M}_T(B) = \mathcal{M}_U(B)$. Suppose that $\mathcal{M}_T(B) = \begin{bmatrix} a_{11} & \cdots & a_{1n} \\ \vdots & & \vdots \\ a_{n1} & \cdots & a_{nn} \end{bmatrix}$. Then we also have $\mathcal{M}_U(B) = \begin{bmatrix} a_{11} & \cdots & a_{1n} \\ \vdots & & \vdots \\ a_{n1} & \cdots & a_{nn} \end{bmatrix}$. Let $j \in \{1, 2, \ldots, n\}$. Then we have

$$T(v_j) = a_{1j}v_1 + a_{2j}v_2 + \cdots + a_{nj}v_n = U(v_j).$$

Since $j \in \{1, 2, \ldots, n\}$ was arbitrary, for all $j = 1, 2, \ldots, n$, $T(v_j) = U(v_j)$. Since B is a basis of V, for all $v \in V$, $T(v) = U(v)$. Therefore, $T = U$. So, F is injective.

Let $\begin{bmatrix} a_{11} & \cdots & a_{1n} \\ \vdots & & \vdots \\ a_{n1} & \cdots & a_{nn} \end{bmatrix} \in M_{nn}^{\mathbb{F}}$. Define $T \in \mathcal{L}(V)$ on B by $T(v_j) = a_{1j}v_1 + a_{2j}v_2 + \cdots + a_{nj}v_n$ for each $j = 1, 2, \ldots, n$. Then clearly $F(T) = \begin{bmatrix} a_{11} & \cdots & a_{1n} \\ \vdots & & \vdots \\ a_{n1} & \cdots & a_{nn} \end{bmatrix}$. So, F is surjective.

Let $T, U \in \mathcal{L}(V)$ and let $a, b \in \mathbb{F}$. Suppose that $F(T) = \begin{bmatrix} a_{11} & \cdots & a_{1n} \\ \vdots & & \vdots \\ a_{n1} & \cdots & a_{nn} \end{bmatrix}$ and $F(U) = \begin{bmatrix} b_{11} & \cdots & b_{1n} \\ \vdots & & \vdots \\ b_{n1} & \cdots & b_{nn} \end{bmatrix}$. Then for each $j = 1, 2, \ldots, n$, we have

$$T(v_j) = a_{1j}v_1 + a_{2j}v_2 + \cdots + a_{nj}v_n \text{ and } U(v_j) = b_{1j}v_1 + b_{2j}v_2 + \cdots + b_{nj}v_n.$$

So, $(aT + bU)(v_j) = (aa_{1j} + bb_{1j})v_1 + \cdots + (aa_{nj} + bb_{nj})v_n$. Therefore,

$$F(aT + bU) = \begin{bmatrix} aa_{11} + bb_{11} & \cdots & aa_{1n} + bb_{1n} \\ \vdots & & \vdots \\ aa_{n1} + bb_{n1} & \cdots & aa_{nn} + bb_{nn} \end{bmatrix} = \begin{bmatrix} aa_{11} & \cdots & aa_{1n} \\ \vdots & & \vdots \\ aa_{n1} & \cdots & aa_{nn} \end{bmatrix} + \begin{bmatrix} bb_{11} & \cdots & bb_{1n} \\ \vdots & & \vdots \\ bb_{n1} & \cdots & bb_{nn} \end{bmatrix}$$

$$= a\begin{bmatrix} a_{11} & \cdots & a_{1n} \\ \vdots & & \vdots \\ a_{n1} & \cdots & a_{nn} \end{bmatrix} + b\begin{bmatrix} b_{11} & \cdots & b_{1n} \\ \vdots & & \vdots \\ b_{n1} & \cdots & b_{nn} \end{bmatrix} = aF(T) + bF(U).$$

So, F is a vector space homomorphism (in other words, F is a linear transformation).

Now, let $T, U \in \mathcal{L}(V)$. Suppose that $F(T) = \begin{bmatrix} a_{11} & \cdots & a_{1n} \\ \vdots & & \vdots \\ a_{n1} & \cdots & a_{nn} \end{bmatrix}$ and $F(U) = \begin{bmatrix} b_{11} & \cdots & b_{1n} \\ \vdots & & \vdots \\ b_{n1} & \cdots & b_{nn} \end{bmatrix}$. Then for each $j = 1, 2, \ldots, n$, $T(v_j) = a_{1j}v_1 + a_{2j}v_2 + \cdots + a_{nj}v_n$, $U(v_j) = b_{1j}v_1 + b_{2j}v_2 + \cdots + b_{nj}v_n$.

We have $(TU)(v_j) = T\big(U(v_j)\big) = T\big(b_{1j}v_1 + b_{2j}v_2 + \cdots + b_{nj}v_n\big) = c_{1j}v_1 + c_{2j}v_2 + \cdots + c_{nj}v_n$, where $c_{ij} = a_{i1}b_{1j} + a_{i2}b_{2j} + \cdots + a_{in}b_{nj}$. Also, we have

$$\begin{bmatrix} a_{11} & \cdots & a_{1n} \\ \vdots & & \vdots \\ a_{n1} & \cdots & a_{nn} \end{bmatrix} \cdot \begin{bmatrix} b_{11} & \cdots & b_{1n} \\ \vdots & & \vdots \\ b_{n1} & \cdots & b_{nn} \end{bmatrix} = \begin{bmatrix} c_{11} & \cdots & c_{1n} \\ \vdots & & \vdots \\ c_{n1} & \cdots & c_{nn} \end{bmatrix},$$ where $c_{ij} = a_{i1}b_{1j} + a_{i2}b_{2j} + \cdots + a_{in}b_{nj}$.

It follows that $F(TU) = \begin{bmatrix} c_{11} & \cdots & c_{1n} \\ \vdots & & \vdots \\ c_{n1} & \cdots & c_{nn} \end{bmatrix} = \begin{bmatrix} a_{11} & \cdots & a_{1n} \\ \vdots & & \vdots \\ a_{n1} & \cdots & a_{nn} \end{bmatrix} \cdot \begin{bmatrix} b_{11} & \cdots & b_{1n} \\ \vdots & & \vdots \\ b_{n1} & \cdots & b_{nn} \end{bmatrix} = F(T) \cdot F(U)$.

Let $i \in \mathcal{L}(V)$ be the identity function, so that $i(v_j) = v_j$ for each $j = 1, 2, \ldots, n$. Then $F(i) = I$, where $I = \begin{bmatrix} 1 & 0 & \cdots & 0 \\ 0 & 1 & \cdots & 0 \\ \vdots & \vdots & \ddots & \vdots \\ 0 & 0 & \cdots & 1 \end{bmatrix}$, the identity for multiplication of $n \times n$ matrices.

Therefore, F is a ring homomorphism. Since F is both a vector space homomorphism and a ring homomorphism, F is a linear algebra homomorphism. Since F is also bijective, F is a linear algebra isomorphism. □

About the Author

Dr. Steve Warner, a New York native, earned his Ph.D. at Rutgers University in Pure Mathematics in May 2001. While a graduate student, Dr. Warner won the TA Teaching Excellence Award.

After Rutgers, Dr. Warner joined the Penn State Mathematics Department as an Assistant Professor and in September 2002, he returned to New York to accept an Assistant Professor position at Hofstra University. By September 2007, Dr. Warner had received tenure and was promoted to Associate Professor. He has taught undergraduate and graduate courses in Precalculus, Calculus, Linear Algebra, Differential Equations, Mathematical Logic, Set Theory, and Abstract Algebra.

From 2003 – 2008, Dr. Warner participated in a five-year NSF grant, "The MSTP Project," to study and improve mathematics and science curriculum in poorly performing junior high schools. He also published several articles in scholarly journals, specifically on Mathematical Logic.

Dr. Warner has nearly two decades of experience in general math tutoring and tutoring for standardized tests such as the SAT, ACT, GRE, GMAT, and AP Calculus exams. He has tutored students both individually and in group settings.

In February 2010 Dr. Warner released his first SAT prep book "The 32 Most Effective SAT Math Strategies," and in 2012 founded Get 800 Test Prep. Since then Dr. Warner has written books for the SAT, ACT, SAT Math Subject Tests, AP Calculus exams, and GRE. In 2018 Dr. Warner released his first pure math book called "Pure Mathematics for Beginners." Since then he has released several more books, each one addressing a specific subject in pure mathematics.

Dr. Steve Warner can be reached at

steve@SATPrepGet800.com

BOOKS BY DR. STEVE WARNER

www.ingramcontent.com/pod-product-compliance
Lightning Source LLC
Chambersburg PA
CBHW081722100526
44591CB00016B/2470